# STELLINGEN.

## I

Noch JANSSEN's chiasmatypie- noch WINKLER's Konversionstheorie bieden eene bevredigende verklaring voor het genetische verschijnsel van ,,crossing-over''. De cytologische verklaring moet in de
eerste plaats gezocht worden in den overgang van de attractie-
krachten en de afstootingskrachten van de homologe chromosomen
om en bij het tijdstip van meiosis.

## II

De quantitatieve theorie van de geslachtsbepaling wordt gesteund
door de feiten waargenomen bij Lebistes reticulatus.

## III

De in 1931 bij Zea Mays gevonden erffactor voor Chocolate pericarp
(Ch) kan om gegronde redenen beschouwd worden als behoorende tot
de 10de kuppelingsgroep.

## IV

Volgens den huidigen stand der cyto-genetica, kan de lineaire orde
van de genen als volkomen bewezen worden beschouwd.

## V

De nieuwere resultaten betreffende translocatie, inversie, ,,defi-
ciency'' en polyploidie moeten bij de theorieën over het ontstaan van
species in aanmerking genomen worden.

## VI

De proeven van BERGNER (1928) over den invloed van verlenging van elke periode in de levenscyclus van Drosophila melanogaster op het proces van ,,crossing-over" beantwoorden niet aan de eischen van wetenschappelijk onderzoek.

## VII

Het relatief hoogere percentage van mortaliteit onder het mannelijke geslacht bij menschen en dieren waargenomen, hangt primair niet samen met het cytologische verschil tusschen het mannelijke en vrouwelijke geslacht, zooals HUXLEY, LENZ en SCHIRMER zich dat voorstellen, maar is van meer secundair karakter.

## VIII

Dat ultraviolet licht geen merkbaren invloed uitoefent op de frequentie van mutatie bij Drosophila melanogaster en wel op de frequentie van ,,crossing-over", is naar mijne meening daaraan te wijten, dat de kiemcellen waarschijnlijk indirect en niet direct door ultraviolet licht beinvloed worden.

## IX

Het voederen van jonge aquarium visschen met infusoriën, is niet aan te bevelen, behalve voor luchthappende visschen.

## X

Voor genetische onderzoekingen van eencellige polyenergide organismen, is de protist Stentor roeseli zeer geschikt.

## XI

Bij de biometrische beoordeeling van de scheefheid eener curve geve men zich rekenschap van het feit, dat de modus zich noodzakelijk moet verplaatsen in de richting van de kleinere varianten bij het vergrooten van de klasseruimtemaat, zoodat eene curve dan pas symmetrisch kan zijn wanneer de kleinste variant(en) even groot of grooter is (zijn) dan de klasseruimtemaat.

## XII

Het verschijnsel van het onderscheidingsminimum en de daarmede samenhangende wet van WEBER moet physiologisch verklaard worden en niet zooals HEYMANS en zijn school dat trachten te doen door middel van een psychische remming.

## XIII

De methoden van PAVLOV in verband met de conditioneering van spieren en klieren behooren de introspective methode in de zintuigs-physiologie te vervangen.

## XIV

Bij de anthropometrie moet men uit den aard der zaak meer varianten hebben voor het bepalen van een gemiddelde dan ge-woonlijk het geval is in de biometrie, waaruit ook volgt, dat paleo-metrische gemiddelden betrekkelijk onbetrouwbaar zijn.

## XV

De ,,crossing-over'' theorie van BAUER aangaande de erfelijkheid der menschelijke bloedgroepen is foutief.

## XVI

Door eene methode te vinden om rachitis aan fossiele beenderen te erkennen kan men een conclusie trekken aangaande de meteoro-logische toestanden van de periode in kwestie.

## XVII

Het zou voor de praehistorische anthropologie van groote waarde
kunnen zijn indien men bij het bloedgroepenonderzoek een pleiotrope
werking van de bloedgroepgenen   op het skelet zou kunnen vast-
stellen, hetzij direct of indirect (spieraanhechting).

## XVIII

De gewoonte van veldbrand in Zuid-Afrika moet wettelijk verboden
en bebossching zooveel mogelijk aangemoedigd worden.

A THEORETICAL AND EXPERIMENTAL STUDY ON THE
CHANGES IN THE CROSSING-OVER VALUE

*(Overdruk uit Genetica XIV Afl. 1—2)*

# A theoretical and experimental study on the changes in the crossing-over value, their causes and meaning

PROEFSCHRIFT TER VERKRIJGING VAN DEN
GRAAD VAN DOCTOR IN DE WIS- EN NATUUR-
KUNDE AAN DE RIJKSUNIVERSITEIT TE GRO-
NINGEN, OP GEZAG VAN DEN RECTOR MAGNI-
FICUS Dr A. G. ROOS, HOOGLEERAAR IN DE FA-
CULTEIT DER LETTEREN EN WIJSBEGEERTE,
TEGEN DE BEDENKINGEN VAN DE FACUL-
TEIT DER WIS- EN NATUURKUNDE IN HET
OPENBAAR TE VERDEDIGEN OP MAANDAG
30 MEI 1932, DES NAMIDDAGS TE $2\frac{1}{2}$ UUR

DOOR

## GERHARDUS ELOFF
GEBOREN TE RUSTENBURG, TRANSVAAL

Springer-Science+Business Media, B.V.

1932

ISBN 978-94-017-5856-7     ISBN 978-94-017-6331-8 (eBook)
DOI 10.1007/978-94-017-6331-8

AAN MIJNE VROUW

Dit is my seer aangenaam om my waardering uit te spreek jeens al diegene wat bygedra het tot my wetenskaplike vorming.

In die eerste plaats U Hoogewaardeerde Promotor, Hooggeleerde TAMMES. Aan U het ek dit te danke dat ek een van my grootste lewensideale kon sien verwesenlik word n.l. die moontlikheid om werksaam te kan wees op die gebied van die Genetica. U suiwere kritiek en warme aanmoediging sowel by my proefneminge as by die opstelling van die dissertasie, word seer hoog op prys gestel. U het verlede jaar deur die organisere van die genetiese ekskursie na Berlyn ons in staat gestel om met vooraanstaande persoonlikhede en ver- maarde institute in aanraking te kom. Ook hierdie voorreg word seer hoog op prys gestel. Maar ook het U steeds belangstellend gewees vir my persoonlike aangeleenthede en in nie te rooskleurige tye was U ons steeds 'n steun. Dit sowel as besondere voorregte deur U aan my toegestaan, word seer gewaardeer.

Dit mag my seker vergun word om daarop te wys dat U as een van die pionier-opbouers van die erflikheidswetenskap, elke tree vooruit- gang daarvan meegemaak het, en U vind dit seker 'n skone beloning om, in aansluiting bij U eksakte natuur, dit vandaag met trots en vreugde te sien as uitgegroei tot die eksakste gebied in die biologie. Temeer voel ek vereer hierom dat ek die voorreg had gevorm te word deur U as 'n pionier-opbouer. My nederige wens is dat U nog lang vir die wetenskap gespaar mag bly.

Ook U Hooggeleerde STOKER en COETZEE is ek veel verskuldig. U entoesiasme en sin vir eksaktheid het 'n groot indruk op my ge- maak.

Ek het die eer gehad Hooggeleerde BRUGMANS om te profiteer van U filosofiese insigte veral as een van die huidige leiers van die Hey- mansskool. Wat ek van U geleer het asook U tegemoetkomings by my studie word hooggewaardeer.

Seer geleerde van GIFFEN, 'n groot voorreg het ek gehad om onder U leiding kennis te maak met die fossiele mens en sy afstammings- vraagstuk. Seer tegemoetkomend was U in die voorsiening van die nodige instrumente vir antropometriese oefeninge.

Hooggeleerde BUYTENDYK, dit was my 'n eer om by U te kon werk en van U kolleges te kon volg. U pionierswerk om op natuurwetenskaplike wyse in te dring in die dieresiel het altyd 'n diep indruk op my gemaak.

Seer vormend vir my was die laboratoriumsfeer geskep deur U Hooggeleerde MOLL, SCHOUTE en ARISZ; ook vir wat ek aan U verskuldig is my opregte dank. Tewens was dit my 'n eer om met U kennis te maak. Hooggeleerde ARISZ, U vriendelike raad en tegemoetkoming met die vir my proewe nodige toestelle word hoog gewaardeerd. Besonder aangenaam ook was dit vir my om met U kennis te maak Hooggeleerde DANSER en Seergeleerde KUIPER.

Waarde WOLTHUIS, U raad en hulp by my Lebistes proewe is zeer gewaardeerd.

Waarde VEENHOFF, met veel sorg het U my termostate sien te vervaardig. My opregte dank daarvoor.

Waarde HOEKSEMA, U het die tekeninge vir my dissertasie met groot sorg en keurigheid vervaardig, waarvoor my opregte waardering.

Waarde VEENHOFF Jr en ALKEMA, ook van U het ek veel geprofiteer. Ontvang my dank daarvoor.

Ek sou hier graag 'n groet wil rig aan al die bioloë met wie ek nader kennis gemaak het aan hierdie Uniwersiteit. Ons omgang met U Seergeleerde DE HAAN, Waarde Mej. KOK, Waarde MEKEL, Waarde Is. DE HAAN, Seergeleerde ALGERA was heel aangenaam en leersaam.

'n Woord van opregte dank aan die Nederlandsch Z. Afrikaansche Vereeniging, die Z. A. Voorschotkas en die Beurse Komitee te Pretoria. Ons benagstellende famielielede en alle andere in Suidafrika wat op een of ander wyse die voltooiïng van my studies in Nederland moontlik gemaak het, word hiermee opreg bedank.

# A THEORETICAL AND EXPERIMENTAL STUDY ON THE CHANGES IN THE CROSSING-OVER VALUE, THEIR CAUSES AND MEANING

by

## G. ELOFF

(With 12 figures)

TABLE OF CONTENTS

# INTRODUCTION

In 1920 DETLEFSEN (29) published an article entitled "Is Crossing-over a function of distance?" Apparently this was not altogether approved of by MORGAN's school, for according to them DETLEFSEN did not quite grasp the true meaning of the chromosome maps, with the consequence that he underrated the research in connection with chromosome topography, for he stated: "We do not know that the distance which gives 1 % (or n %) of crossing-over is a fixed unit, . . . . our arbitrary unit of measurement may itself prove to be a variable". He (DETLEFSEN) pointed out that he succeeded through selection experiments to reduce crossing-over in a certain case from 33 % to 0 %, in view of which consideration it would perhaps be simpler to conclude that linkage is not a function of distance, i.e. crossing-over is not necessarily proportional to distance. DETLEFSEN's criticism was immediately answered by STURTEVANT and others: "One unfamiliar with the literature of the subject would probably infer from DETLEFSEN's paper that the possibility of inherited linkage variations had not been taken into account by those concerned in constructing chromosome maps. In point of fact, the matter has not only been taken into account, but has often been discussed in the literature . . . .". This in fact is true for already in 1919 MORGAN (78, p. 139) wrote: "Crossing-over is not absolutely fixed in amount but is variable. This statement does not refer to variability in the number of crossing-over due to random sampling but to fluctuations in environmental conditions, or due to internal changes in the mechanism of crossing-over itself". Thus STURTEVANT also quoted several cases from which it was apparent that the variations in linkage were duly considered in connection with the construction of chromosome maps. But he also pointed out that these maps were intended to show the actual sequence of the loci and the relative amount of crossing-

over between them. The intervals between adjacent loci are not to
be taken as necessarily proportional to the actual spatial distances
between them, though the distance is evidently one of the elements
concerned. In view of these considerations STURTEVANT is of opinion
that it is clear that DETLEFSEN has misunderstood the significance
of the published maps.

To my mind STURTEVANT in his turn went too far in underrating
the distance factor. Today however he might probably think differently about this matter, for cytogenetics has clearly proved that
distance is not only one of the elements concerned, but that crossing-
over is essentially a function of distance after all. Still more, its variability has proved itself to be a valuable instrument to detect inter-
and intra-chromosomal differences especially when studied in co-
operation with cytology. There are great future possibilities for
cytogenetic research in connection with translocation, non-disjunction and related phenomena. Research in this direction although of
very recent date has resulted in far-reaching discoveries, thanks to
the artificial production of these abnormalities by means of X-rays.
Through studies on translocation we came to learn that genetic
distances may widely differ from cytological chromosomal distances,
from region to region. This of course means that the actual chromosome differs physically from region to region. This also is how we came
to know that the mid-region of the second chromosome is essentially
different from regions either to the right or the left thereof, in being
far more responsive to environmental influences. This region now we
know is the place of attachment of the spindle-fibre.

It is clear therefore that the variability of the crossover percentage
(c.o.v.) [1] should not only be studied for its own sake but also with
the intention of obtaining the fullest possible data of inter-and intra-
chromosomal differences, so that possibly in co-operation with experimentally evoked translocations, non-disjunctions etc., we may ultimately be able to construct a map more in agreement with the chromosome as seen through the microscope, that is to say with the loci
of the genes more in correspondence with the cytological loci.

Research in this direction involves the use of multiple recessive
mutants as far as possible distributed over the whole length of the

---

[1]  In this dissertation I intend substituting c.o.v. for crossover percentage and
for crossing-over value.

chromosome in question, but it is also evident that as far as possible those factors causing the changes in the c. o. v. should be known and duly accounted for. In the following pages the factors so far discovered in this connection, will be dealt with. Environmental influences it is clear, may be controlled relatively easily, but it would be a much more difficult task to account for those changes in the c.o.v. which are due to genetic causes e. g. crossover modifiers. We therefore thought it fit to devote more space to the chapters dealing with crossover modifiers and other genetical influences as for instance difference in crossing-over among male and female gametes not only for *Drosophila* but also for other objects studied.

In Part II of this dissertation our own experiments are dealt with, but the main idea underlying both Part I and Part II is that we have entered a new era in cyto-genetics which offers valuable means to construct true chromosome maps. This is a great task but if it is attained, it will be one of the greatest triumphs in biology.

The experiments dealt with in Part II were made in the Genetic Institute of the State University of Groningen. The invaluable criticism of Prof. Dr. TAMMES under whose supervision I had the honour to carry out this investigation, is highly appreciated.

# PART I

## CHAPTER I

### THE EFFECT OF TEMPERATURE ON CROSSING-OVER

The classical experiments of PLOUGH showing the far-reaching effect of temperature on the phenomenon of crossing-over in *Drosophila melanogaster* convinced geneticists that physical and physiological agents may cause changes in the linkage values. As a consequence, a train of experiments were commenced with in this direction, with a view of obtaining new insight and understanding with regard to the mechanism of crossing-over and especially as regards its variability. PLOUGH's experiments however, involved other factors besides temperature, e. g. age, food and the very important phenomenon of inter-chromosomal differences in susceptibility to these factors.

PLOUGH (88) observed that there existed a remarkable difference in crossing-over between cultures of otherwise genetical identity. These cultures might even be raised from the same mother, they might be made up in the same manner, from the same stock, and in the same room. But there probably was a difference in the temperature conditions e. g. 20° C and 25° C. It was a most remarkable fact that PLOUGH happened to experiment on the black (b) — purple (pr) — curved (c) region of the second chromosome, for as will be pointed out this central region seems to be relatively more liable to changes in crossing-over than other regions. The different crossover values obtained for the same region in question were statistically significant and could not be ascribed to chance distribution.

He (88) systematically studied the correlation between the crossover percentage and temperature, the results of which were expressed in his wellknown curve showing the influence of different temperatures on crossing-over between b-pr. See Fig. 1.

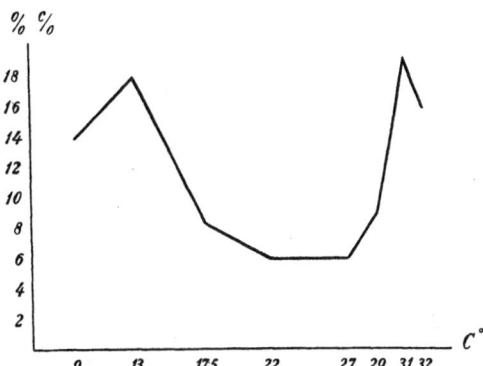

Fig. 1. Curve showing influence of different temperatures on crossing-over between black and purple
in the second chromosome. After Plough.

Temperatures between 19° and 27° C showed no significant
differences for the region black-purple. This is probably the
reason why *Droso-phila* cultures raised under conditions of
ordinary room temperatures very seldom show deviating results.
From the curve it is seen that the rise in crossing-over reaches its
first maximum at 13° C, and a second and still higher maximum at
31° C. PLOUGH called attention to the physiological similarity
between his curve and that shown by the amount of contraction
of a frog's muscle.

No temperature influences on crossing-over could be detected in
connection with chromosome I and chromosome III. For chromosome
I were involved the regions vermilion, sable, garnet, forked, (v-s-g-f).
In PLOUGH's own words we emphasize this inter-chromosomal difference in reaction: "Whatever is the meaning of this difference in
reaction to temperature, it gives one added reason for believing that
the "genetic chromosomes" are discreet elements which differ among
themselves and retain this individuality from generation to generation".

Temperature influences furthermore proved to be a suitable means
to find the probable stage during which crossing-over takes place.
Extreme temperature treatment of *Drosophila* females during the
early stages of development or during the late pupa stage shows a
significant rise in the c. o. v. for the first, but not for the second brood.

When a full-grown female is thus treated, the effect on crossing-over is not noticed untill 225—275 eggs are laid. It is therefore clear that temperature affects crossing-over at a certain point during the oögenesis, and further evidence makes it probable that crossing-over takes place during the very earliest oöcyte. This agrees with the fact that temperature influences have a great effect when applied to *Drosophila* in the early larva stage.

Temperature influences however, may be ineffective in the presence of other genetic factors. For instance, a case was met with where extreme temperature had no effect on crossing-over in the black-purple region of the second chromosome of *Drosophila melanogaster*. This was due to the presence of the crossover modifier C III found by STURTEVANT. It also anihilates the effect of temperature with regard to the regions Star-black and black-curved. Also according to the investigations of other geneticists the central regions of the V-shaped autosomes proved to be particulary sensitive to changes in crossing-over percentages due to external agencies. With regard to chromosome I no effects were detected by PLOUGH. But this will become clear when it is taken into consideration that only later STURTEVANT located new genes to the right of the X-chromosome. ANDERSON, BRIDGES, MORGAN and STURTEVANT (105) are of opinion that this right hand end of the X-chromosome corresponds with the point of attachment of the spindle-fibre. This region may therefore also prove itself to be sensitive to temperature and other influences. This offers new field for investigation. The result was that STERN (105) planned extensive experiments to investigate the sensitivity of the crossing-over values of chromosome I under external influences. He too made use of extreme temperatures, and furthermore he treated females of different ages. He studied the region from garnet at 44.8 units to bobbed at 70, which loci are nearest to the place of attachment of the spindle-fibre. The results obtained with regard to the Bar-bobbed region were as follows: Females $\dfrac{B+}{+bb}$ were crossed with bb males. Some of the females were kept at 25° C as control flies, others hatched and crossed at 30° C, were first kept at 30° C and thereafter transferred to 25° C. Two main points were observed. 1. The females raised at 30° C gave a much higher c.o.v. during the first 3 to 4 periods of transfer than the controls. Differences in recombination between cultures kept at 25° C (controls) and cultures kept at 30° C were statistically significant, being $8.5 \times$ the P. E. 2. The first minimum in crossing-over percentage was observed at the age of 4 or 5 days of the female for both temperatures. The crossing-over difference between females 3—7 days old and 7—13 days old was significant.

Apart from the studies on *Drosophila* I could trace only one more case where the influence of temperature on crossing-over was studied. It concerns the experiments of CASTLE and WACHTER (22) on rats and mice. Their results were negative. I do not think however, that these experiments really are of direct bearing on the question, because it was in fact, seasonal influences that were studied. Furthermore, seasonal influences involve so many factors as only to complicate the results. It is therefore perhaps the safest to say that exact data concerning temperature influences on crossing-over in rats and mice and in other mammals are not on hand.

Nor could I trace any experimental data with regard to plants, although we are sure that the experiments of HIORTH with artificial illumination offer great facilities to control the external influences on plants, which are otherwise so much exposed to meteorological influences.

It would have been very interesting to know whether the process of crossing-over was influenced directly or indirectly by temperature. Unfortunately the experiments discussed offer no clue.

## CHAPTER II

### The effect of X-rays and of radium on crossing-over

The introduction of radium and X-ray treatment for experimental methods in genetics has been of very great advantage to this science and it will probably still be so for many years to come. So far I could only trace experiments with *Drosophila* in connection with the influence of radium and of X-rays on the phenomenon of crossing-over. Again the investigations concerned the first and the second chromosomes. MAVOR (70) began these experiments on the first chromosome and particularly on the section we-m (eosine-miniature). A white-eyed normal-winged female was crossed with an eosine-miniature male. For each experiment a $F_1$ female $\dfrac{w \quad M}{we \quad m}$ was X-rayed with an ener-

gy of 50,000 volts, the doses being varied as follows: 21D, 23D, 26D, .... 35D and 37D, where $D = \dfrac{\text{milliamperes} \times \text{minutes}}{(\text{distance in decm})^2}$ .
A sister female hatched at the same time was kept as a control. The treated as well as the not-treated female was kept in the first bottles for six days and in the second bottles for eight days. The offspring were counted untill the eighteenth day from the day the parents were put in the bottles. No significant difference in crossing-over could be detected among the first bottles neither with regard to the different X-ray doses nor as far as the total data were concerned. As regards the second bottles however, a considerable reduction in crossing-over was observed for the treated females. It was furthermore clear that this reduction was the more remarkable the stronger the doses were, for instance the c.o.v.'s obtained with the doses 21D—29D differed from that of 35D—49D by 4.8 $\times$ P. E. diff.

Another series (MAVOR's 4th) of experiments was put up more or less in the same manner as the above mentioned, except for (1) the fact the doses were 35D—38D, which in another series of experiments proved to be the most suitable doses for modifying the c.o.v., and (2) differences in duration of X-ray treatment which were as follows:

group 1 received 35D for 3 minutes and 17 seconds,
„ 2 „ 38D „ 2 hours and 15 minutes,
„ 3 „ 36D „ 20 hours and 20 minutes.

Group 3 therefore received more radiant energy as well as approx. 400 times longer treatment than group 1. Both the treated and the control females were transferred to new bottles every three days, and the offspring were counted up till the eighteenth day after the parents were put in the bottles. No significant difference was observed between the c.o.v.'s of the X-rayed and the control females during the first six days. Hereafter however a remarkable difference was observed for the third and fourth bottles, (7th—12th day). For the three groups together the third bottles gave 9.32 % of crossing-over while the control gave 27.4 % crossing-over the difference of which was 12.85 $\times$ the P. E. diff. For the fourth bottles 9.8 % of crossing-over was obtained for the X-rayed flies as against 28.7 % for the controls, a difference of 13.6 $\times$ P. E. diff. The X-ray effect was thus observed during the second period of 6 days. But group 1 showed the same result apart from the others. It seems therefore, that it is not the duration, but the total radiant energy that matters.

Further analysis of the data seems to point out that crossing-over modification must have happened more or less at the time when the flies began to regain their partially lost or disturbed fertility, which was caused by the X-ray treatment.

By adding together the data obtained from series 3 and 4, (on the ground that the females were of the same genetic constitution and kept under similar conditions) it was found that the difference of crossing-over between the treated and the control females for the first 6 days amounts to but 2.03 × P. E. diff. For the eggs produced for the second 6 days however, the difference was 28.37 × P. E. diff.

The evidence, obtained was insufficient to make out whether these modified c.o.v.'s induced by X-ray treatment, were inherited. But the data available point to the contrary.

These results, it will be remembered, concern the experiments on the eosin-miniature section of the 1st chromosome.

For the purple-curved region of the 2nd chromosome it is interesting to note that the crossing-over difference for X-rayed as well as for females treated by temperature, as compared with the controls, became apparent on the 7th day, and reached its maximum on the 8th day. From then it decreased so that on the 9th day it stood as on the 7th. Hereafter the c.o.v. for the X-rayed females behaved differently from that of the temperature treated females. Henceforth the c.o.v. of the temperature experiment remained the same as that of the controls, while that of the X-rayed females remained differently from the c.o.v. of the controls up till the 13th day. Crossing-over values on the 14th and 15th days showed no significant difference between c.o.v.'s for X-rayed and controls compared. It may be added that the X-ray dose of 32D showed a greater effect on crossing-over than the temperature treatment of 30° C, that is to say for the region black-purple.

MAVOR and SVENSON (74) continued the research in this direction on the 2nd chromosome of *Drosophila melanogaster*. They chose the same region, black to curved, studied by PLOUGH in connection with temperature. These c.o.v.'s, in the meantime, for the region black to curved had been standardized by MULLER, BRIDGES and PLOUGH. Crossing-over between black and purple was found to be 6.2 %, between purple and curved, 19.9 %. The percentages of the different investigators were more or less the same, the number of flies counted were 50.000 and 60.000 respectively.

The experiments of MAVOR and SVENSON may be specially mentioned as experiments with the intention to throw more light on the physiology of the changes in the c.o.v. They wanted to find out whether the observed changes in linkage were due to direct changes in the mechanism of crossing-over as such or whether it was a result of the general physiological condition, which caused among others changes in crossing-over. In this connection for instance it was of great importance to study the relative viability of the different classes.

Out of 20 heterozygous female flies, 11 were kept as controls, and 9 were treated with X-rays for 3 minutes 15 seconds at a distance of 25.5 cm from the tungsten target; the Coolidge tube was worked at 50.000 volts and .05 amperes. A day after treatment all the females were back-crossed to black-purple-curved males. The bottles were changed every 3 days up till the 18th day when all the females were killed. The results obtained showed that the difference in crossing-over of the 2nd and especially of the 3rd bottles as compared with the controls was certainly significant for the region black-purple. The same applies to the region purple-curved, but to a lesser degree.

A second experiment was performed with 11 control and 27 X-rayed females, the treatment this time being exercised for three minutes. The females were killed on the 12th day and the counts were done every three or four days for 17 days after mating; the temperature was 22° C in both cases. The first bottles showed no difference in crossing-over, while the second and the fourth but especially the third bottles showed a rise in crossing-over for the region black-purple. As far as the region purple-curved is concerned, a considerable rise in crossing-over was detected for the second and the third bottles, while no statistically significant difference was observed for the fourth bottles.

This second experiment was in general agreement with the first. The combined results affirmed the fact that X-rays had a similar effect on crossing-over in both regions, with this difference however that for the region purple-curved the rise is relatively smaller than for the region black-purple. Also the "recovery" till the normal c.o.v. is reached, is attained sooner for the purple-curved region.

Thus these experiments on the second chromosome offer a clear case of opposite effect of the same physical agent. It will be remembered that in the case of the first chromosome crossing-over between eosine and miniature showed a decrease in crossing-over and the more so the stronger the X-ray dose was. It may be recalled that PLOUGH obtained a rise in crossing-over in connection with extreme temperature influences on crossing-over for the same regions b-pr-c. It seemed as if the effect of X-raying was displayed sooner and lasted longer than the effect of temperature. Later experiments however did not agree with these findings. The effect of temperature disappeared suddenly while its duration corresponded with that of the treatment, whereas the effect of three minutes of X-ray treatment lasted until about the 15th day. As a matter of fact temperature affected crossing-over only at a certain stage during the development of the eggs. X-raying of *Drosophila* females before mating or directly after hatching soon showed a rise in crossing-over for the black-purple region. The mechanism of crossing-over therefore was not affected by X-rays at a certain point of time during the development of the egg so that the process of crossing-over seemed to be affected rather indirectly. It might be that a general condition was created which on its turn produced a further effect on the mechanism of crossing-over. Strangely enough this was contradicted by later experiments.

In 1924 MAVOR and SVENSON (73) published the results of experiments which made possible a direct comparison of the effect of X-rays and of temperature on linkage in the 2nd chromosome of *Drosophila*. These results were better suited for comparison than those of the former experiments, since the controls as well as the temperature-treated and X-rayed females were sisters. The treatment was commenced at the same time; controls and X-rayed females were constantly kept at a temperature of 23° C. The temperature treatment lasted for 48 hours, but after the treatment these females also were kept constantly at 23° C. Again the regions concerned were black-purple-curved. The X-rays doses were 32D, so that in these experiments the dose was somewhat stronger than in that of their former, in which case it was 29.4D. The accuracy was furthermore enhanced by using 1 day bottles (24 hrs). The flies were kept in the 1st two bottles for two days but in the 3rd till the seventh bottles they were kept for one day, for the 8th and the 9th for 2 days.

The effect of X-rays and temperature became evident at the same time, treatment having been applied at the same time. According to the former experiments, the effect of X-rays showed itself sooner than that of temperature. (MAVOR and SVENSON (74). It might be that some technical differences were responsible for these contrary results. In the latter experiment both physical agents again caused a rise in crossing-over for both regions concerned. The effects became evident in the $F_1$ derived from eggs laid on the 6th and the 7th day after the treatment. It appeared that temperature effects became apparent earlier, as far as the rise in crossing-over was concerned, for the black-purple region, where the change became evident on the 6th day, as compared with the purple-curved region. With regard to the duration of the effect it was observed that an X-ray treatment for only 20 minutes nevertheless lasted for a considerable time.

In connection with temperature treatment on the black-purple region it was observed that the difference between the c.o.v.'s of the control and the treated females disappeared on about the 9th day. In contrast to this, X-ray effect lasted till the end of the experiment. The difference in the c.o.v.'s of the control and X-rayed flies was still 10.34 P. E. diff. for the 14th and the 15th day for their experiment No. 406.

A few years later, MAVOR (71) faced the question "whether induction of non-disjunction and the modification of the c.o.v. is due to a direct effect of the X-rays on the germ cells or to a general effect of the X-rays on the physiological condition of the fly". Extensive experiments were started to study the effect of X-raying the *Drosophila* pupae in different ways. In some cases the whole pupa, in others either the posterior or anterior part was X-rayed, and protection of that part which was not to be X-rayed was effected by means of silver bars. It was found that the c.o.v. was only modified when the posterior part of the pupa or when the whole pupa was X-rayed. The results were conclusive, proving that crossing-over was only modified by X-rays when the germ cells are affected directly. It was therefore not due to the general physiological effect of X-rays on the pupa.

Before discussing the problem of the regionally differential effect of X-rays on crossing-over in the chromosomes of *Drosophila* it may be well briefly to mention the influence of radium radiations on crossing-over. In this connection we might recall the experiments of PLOUGH (91), with pure bromide treatment. Once more the experiments concerned the regions black-purple-curved of the 2nd chromosome. Five newly hatched heterozygous females were treated, discontinuously, the intervals having been different for the different flies. Five sisters were not treated and were kept as controls. These as well as the treated flies were back-crossed to black-purple-curved males. The bottles were changed every four days, and the counts were done for twelve consecutive days.

From the results it was clear that radium radiations caused a general rise in the c.o.v. which was the more striking for the larger dose. After 40 minutes of exposure to radium radiations the effect was apparent between the 5th and the 8th day, but reached its maximum between the 9th and the 12th day. Females exposed for 20 minutes on two consecutive days showed a possible increase in crossing-over between the 5th—8th day. The treatment which lasted 20 minutes showed a small decrease in the c.o.v. for the two first periods of 4 days. PLOUGH concluded that the shorter treatment (20 min) caused a decrease of crossing-over in eggs nearest to the point in time when crossing-over takes place, while it caused a small rise in eggs still further back in the early stage of oögenesis.

MAVOR and SVENSON's (70, 74) discovery that the same dose of X-rays caused a rise of crossing-over in the 2nd chromosome but a decrease in the X-chromosome might be explained by inter-chromosomal differences, supporting the theory of the individuality of the chromosomes and their corresponding linkage groups. But it must be admitted that it is quite different from the case where a small difference in the amount of the same physical agent caused an opposite effect on the c.o.v. This brings us close to the phenomenon of intrachromosomal differences in crossing-over.

It will be remembered that X-rays had a more striking effect on the region purple-curved of the 2nd chromosome than on its shorter region black-purple. It may be remarked that temperature exercised just the opposite influence in so far as this shorter region reacted more strongly than the longer region purple-curved. Considering these phenomena MULLER (79) put himself the question whether still other regions of the 2nd chromosome may possibly show even greater differences in crossing-over in connection with X-ray treatment. He was logically led to this investigation because of the following genetic and cytological considerations.

Cytologically speaking it is a striking fact that the central regions of both the 2nd and the 3rd chromosomes (V shaped) differ from the distal regions in 4 important respects: 1. the spindle-fibre is attached to the central regions and not to the distal regions; 2. the bending of the chromosomes is located more or less in the central regions; 3. the chromatine seems to stain less deeply in these central regions; 4. the chromosome is narrowed in the neighbourhood of this region.

Genetically, 1. PLOUGH's experiments (1917, 1924) showed that extreme temperature affected crossing-over in the central region and not in the distal regions; 2. BRIDGES and MORGAN (1915, 1919, 1923) observed more or less the same phenomenon in regard to age effects; 3. MULLER and BRIDGES have found coincidence of crossing-over in the central region to be greater than in the distal regions; 4. the standard map shows a crowding of mutant genes and 5. of lethal genes in this central region. MULLER is of opinion that this apparent crowding may be due to a lower frequency of crossing-over in this region.

In order, as far as possible, to cover the whole chromosome, MULLER made use of multiple stocks. Besides, he varied the X-ray dose for the same region. For the 2nd chromosome he used a stock: dumpy (Td), black (b), purple (pr), curved (c), plexus (px), speck (sp), which stock was briefly named II-ple. A similar III-ple stock was raised for chromosome III, with the constitution roughoid (ru), hairy (h), scarlet (st), pink (p), spineless (ss), and ebony (e).

MULLER's results were briefly as follows: a relatively smaller dose of X-rays (26.8 Holzknecht units) caused a statistically significant rise in the c.o.v. for the central regions of the 2nd and the 3rd chromosomes. This central region measures about 6 units. Neighbouring regions as well as the distal regions were not, or insignificantly affected. Twice this dose of X-rays however caused a significant rise in crossing-over in the 3rd chromosome, not only in the central region but also in the neighbouring regions where the difference between the crossing-over disturbance caused by the smaller doses X-rays and by the heavier was most marked. More distal regions were not affected apparently and it might even be possible that the stronger dose caused a reduction in crossing-over in these distal regions.

With regard to the difference in effect on crossing-over of lighter and heavier doses in the 2nd chromosome, it will be sufficient to compare the results of MULLER's experiments with lighter doses of X-rays with those of MAVOR with heavier doses. It pertains to the region black-purple-curved. Again it was apparent that the heavier dose caused a relatively greater rise in the c.o.v.'s. It also brought about a statistically significant effect on the neighbouring regions. Unfortunately sufficient data concerning the distal ends were not available.

Nevertheless, these results in aggregate clearly prove the existence of intra-regional differences in susceptibility to X-ray effect on crossing-over, in both the 2nd and the 3rd chromosome. The maximum of susceptibility is observed in the bend of these two V-shaped autosomes. "These intra-chromosomal differences in susceptibility are very great, being comparable to the inter-chromosomal differences found by MAVOR between certain regions in the 1st and the 2nd chromosomes" (MULLER, 79).

Even in the presence of a crossover inhibiting factor located in the 2nd chromosome, (WARD's "curly-winged" stock) X-rays nevertheless caused a rise in the c.o.v.'s, and it is possible that the same may be applicable not only to the central region but also to the distal region.

MULLER's experiments were carried out with great statistical care and are therefore of exceptional value. Unfortunately however MULLER did not try to give an explanation of the facts, although strangely enough he was urged on to these investigations by well defined genetical and cytological considerations, and it is only logical to expect, in conclusion, a weighing off against the facts found. To the contrary, the description of technique and statistics occupy a relatively large space.

As far as the first chromosome is concerned it will be remembered that STERN (105) found the right end region from garnet to bobbed (place of attachment of the spindle-fibre) to be sensitive to temperature influences, while PLOUGH found no susceptibility for other regions of the X-chromosome to temperature influences. MAVOR (70) observed reduction of crossing-over between eosine and miniature in the case of X-ray treatment.

## CHAPTER III

CROSSING-OVER MODIFIERS AND OTHER GENETICAL CHANGES IN THE
CROSSING-OVER VALUE

### § 1. *Drosophila melanogaster*

This is probably one of the most fascinating chapters in analytic genetics, and one of the most important chapters in cyto-genetics. The fact is, as RASMUSSON (93) puts it, ".... the linkage values are inherited as any quantitative character of any organism propagating in a sexual way .... We must also conclude that it is most probable that the variations in linkage values represent a quite common and ordinary trait of heredity". It is however absolutely necessary not to underrate the modifications caused by environmental conditions. This RASMUSSON unfortunately did. He stated that all results indicate that modifications caused by environment are almost unimportant. How RASMUSSON possibly could have said this even in 1927—'28 when so many experiments have convinced geneticists of the very great importance of environmental influences on crossing-over, cannot be understood. It might be that due attention to STURTEVANT's (114) article: ,,A case of rearrangement of genes in *Drosophila*" would have mitigated an extreme opinion as RASMUSSON's. This will be mentioned in due course. In the meantime the great importance of crossing-over modifiers and related phenomena e.g. of genetically changed linkage values is admitted, and we may pass on to consider the facts so far obtained.

As a suitable introduction we may briefly discuss the most important facts stated in STURTEVANT's (112) article, "Inherited linkage variations in the 2nd chromosome of *Drosophila*". It concerns the case of a female *Drosophila* from a Nova Scotia stock, in which two genes were detected, which when present in heterozygous condition caused a reduction of the c.o.v.'s in the neighbourhood. One of these genes was found to be inactive when the fly was homozygous for it, a phenomenon well worth considering. These modifiers were detected in the following manner. According to BRIDGES and MORGAN the c.o.v. between vestigial and speck amounts to 37 % approximately. A $F_1$ female from a cross of a wild Nova Scotia female with a vestigial speck male from the laboratory cultures, was back-crossed to vestigial speck males from the latter named stock. Fifty five wild and 44 vestigial specks were obtained but no crossovers. Two females from the 55 wild flies were back-crossed to vestigial speck brothers. From the one female 136 offspring were obtained, 2 of which were crossovers; also, from the other female 120 offspring were obtained, 2 of which were crossovers. Several experiments hereafter confirmed the fact that when a Nova Scotia chromosome is present crossing-over is greatly reduced. These crossover modifiers could be located; CIIl somewhere to the left of purple reduces crossing-over between star and purple in females heterozygous for it; CIIr, located between purple and speck, reduces crossing-over in that region, when present in heterozygous condition. Location is done by back-crossing females heterozygous for different regions of the Nova Scotia chromosome. For instance to topograph CIIr, a female with an original Nova Scotia chromosome, and a black-curved-speck chromosome, thus:

$$\frac{\text{CIIl} \quad + \quad \text{CIIr}}{\text{b} \quad \text{c} \quad \text{sp}}$$ was mated to a black-curved-speck male. A

black female produced by crossing-over would be doubly recessive for black, would not possess the Nova Scotia section to the left of curved, but would have the right end section of this Nova Scotia chromosome, thus: $\frac{\text{b} + \text{CIIr}}{\text{b c sp}}$. This female was back-crossed and

produced 146 offspring without a single case of crossing-over between curved and speck. A wild daughter, $\frac{\text{b} + \text{CIIr}}{+ \text{c sp}}$ was also mated to a

black-curved-speck male and produced 105 offspring including 3 cases of crossing-over between black and curved. Several other experiments were done with the same general result. Thus the cross-over modifier is located in the right hand section of the Nova Scotia chromosome.

As far as the 3rd chromosome is concerned experimental data point in the direction that this Nova Scotia chromosome bears a dominant factor which causes a rise in crossing-over between purple and curved in the second chromosome. This gene called CIII, II, when heterozygous also reduces crossing-over in chromosome III.

It must however be kept in mind that STURTEVANT (114) has changed his opinion of CIII and CIIr or similar crossover-modifiers. He calls attention to the fact that the sequence of the loci is different for different *Drosophila* species, e.g., *D. melanogaster* and *D. simulans*, while data obtained by LANCEFIELD point to a similar rearrangement of genes in *D. obscura*. STURTEVANT emphasizes the similarity between these findings and those in connection with the modifiers CIIr and CIIl. "These 'genes' both cause in individuals heterozygous for them, the disappearance of crossing-over in the immediate regions where the 'genes' themselves lie, and a considerable reduction of crossing-over in neighbouring regions. In individuals homozygous for either of these 'genes' however, the % of crossing-over rises to or beyond that found in normal individuals. Experiments are now under way in an attempt to determine if these 'genes' are really simply inverted chromosome sections, but it will probably be a long task to definitely settle the matter".

WARD (125) found two dominant crossover modifiers which under normal temperature conditions prevent all crossing-over in the 2nd chromosome of his curly-winged stock. When however, the temperature is more or less 30° C, crossing-over occurs freely in the right half of the 2nd chromosome.

The previous year MARIE and JOHN GOWEN (50) published a more or less similar case. Apart from selection, there was found a pair among the offspring of which no crossing-over could be detected as regards the 2nd chromosome for the regions scute-forked notwithstanding the fact that echinus, cut, vermilion and garnet are all located between these two extreme points. In 1922 this stock had passed through no less than 80 generations involving more than 3000 matings without crossing-over having taken place in any of the known regions of the sex chromosome. Further experiments showed that complete linkage in this chromosome is accompanied by the same phenomenon for the black-purple region of chromosome II. Furthermore, it was observed that crossing-over is wholly prevented also in chromosome III for the region Dichaete-hairy, when crossing-over fails for the regions above mentioned in chromosome I and II. This far-reaching effect is probably due to a crossover inhibiting factor. The case is of great theoretical importance. For in the majority of cases such a crossover modifier generally affects crossing-over in its neighbourhood only or at its most in the same chromosome in which it is located. The inhibitor in question affects crossing-over in the first, the second and the third chromosome. It resembles the case of the gene CIII, II discussed above.

The case next in point of time concerns crossover modifiers described by PAYNE (87).

After an artificial selection through 38 generations PAYNE succeeded to obtain a stock in which the usual number of 4 bristles on the scutellum of *Drosophila melanogaster* was raised to 9. But he also observed, that when this 9-bristle stock was crossed to a multiple recessive stock, sepia, spineless, kidney, sooty and rough (se, ss, k, es, ro) all of which are located in the 3rd chromosome, and allowed inbreeding of the $F_1$, a great reduction of crossing-over took place for these regions. It is important to note that so far these crossover modifiers nearly always caused a reduction in crossing-over.

The relatively strong linkage, in PAYNE's opinion was apparently due to a gene present in the 3rd chromosome of his selected stock, this gene being non-lethal in homozygous condition. BRIDGES called this crossover modifier CIIIP.

PAYNE did further experiments in this direction working with a strain having ,,lance-wings''. In one of the 3rd chromosomes he could locate two crossover modifiers, lCIIIPL and lCIIIPR, l = lance, C = c. o. modifier; III = chromosome III; P = PAYNE; L = left; R = right), which suppressed all crossing-over in this chromosome.

Similar, though less striking disturbances in the c.o.v.'s could be explained by assuming a crossover modifier to be present in the other 3rd chromosome from the IIIple stock. This modifier however, had in fact a significant effect on crossing-over only between rough and hairy, for here the standard c.o.v. of 26 was reduced to 18.9 %. It had no effect on the first and second chromosome.

STURTEVANT (1913) reported the case of his CIII. The same factor or one with similar effect and allelomorphic to it was since found by MULLER (1916) on spread and beaded, and by BRIDGES (1923) with a similar effect on maroon-dwarf. CIII, present in heterozygous condition had its strongest effect in the region of sooty(es). No crossing-over was observed between es-ro; between ss-es crossing-over was observed in 2 % of the cases. Nearly no standard c.o.v. was observed between spineless and the regions in the left half of the chromosome. When CIII was present in homozygous condition crossing-over took place normally for both ends of the chromosome.

BRIDGES and MORGAN discovered a crossover modifier CIIIM

which reduced the c.o.v.'s in the regions of the 3rd chromosome as follows: crossing-over for se-ss was 28.1 %; for ss-es it was 0.0 %; for es-ro 0.5 % and for ro-M (Minuta) 0.5 %.

It is a pity that CIIIM was neither studied in its homozygous condition nor with regard to a possible identity of locus with CIII described above, which modifier had more or less the same effect. However, the crossover-modifier CIII St, detected by BRIDGES and MORGAN, apparently caused a rise in the c.o.v. for hairy and scarlet located at 26.5 and 44.0 respectively, thus 17.5 units apart according to the standard map. When CIIISt was present the c.o.v. was raised to 22 %.

GOWEN (48) made an instructive comparison of the variability coefficients obtained by several investigators for various characters of different objects. The extraordinary high coefficient of variability of crossing-over and especially of double crossing-over was very striking and a closer examination was well justified. GOWEN first considered genetical causes, namely the possible effects on crossing-over of changes in the genes between two fixed points. He referred to the factors found by STURTEVANT and MULLER in the V-shaped autosomes causing a reduction in crossing-over. Naturally he asked the question whether not it was possible that all the genes might exercise an influence on crossing-over. In other words it may be that the c.o.v. is the collective result of all the genes present. Furthermore the problem is faced namely, is the effect on crossing-over just as well a function of a gene as is the case of eye colour, body colour etc.

GOWEN applied the method of substituting different allelomorphs, thus obtaining the changes in crossing-over for different constitutions of the heterozygous female. Thus we see that table 5 and 6 (48) differ from each other in that the females were of the consitutions:

$$5. \frac{\text{se} + \text{ss es ro}}{+ \text{D}' + + +}; 6. \frac{\text{se ss} + \text{es ro}}{+ + \text{H}' + +}. \text{Also the were there constitutions:}$$

$$7. \frac{\text{D}' \text{pp ss es ro}}{+ \text{wild}} \quad \text{and} \quad 8. \frac{\text{se cu ss} + \text{es ro}}{+ + + \text{H}' + +}. \quad (\text{D}' = \text{D} = \text{dichaete};$$

cu = curly; pp = peach; ss = spineless; H' = H = hairless; es = sooty; ro = rough).

The results were similar to those of former experiments in this direction: differences in genetic constitution of homologous chromosomes bring about differences in the c.o.v.'s. As was remarked above, GOWEN put himself the question whether or not the c.o.v. was influenced by other or perhaps by all genes collectively. He intended to examine this by means of a selection experiment, and started selection in the direction of the lowest and of the highest c.o.v.'s of *Drosophila* females. As regards low crossing-over he performed selection for six generations applying strictly brother-sister mating. Males, homozygous for sepia, spineless, kidney, sooty and rough, and females heterozygous for dichaete, sepia, spineless, kidney, sooty and rough were mated.

TABLE 1. Results of crossing-over selection experiment. After GOWEN.

| generat. | 1 | 2 | 3 | 4 | 5 |
|---|---|---|---|---|---|
| M.c.o.v. | $37.97 \pm 0.97$ | $47.81 \pm 1.274$ | $49.67 \pm 1.929$ | $49.03 \pm 1.272$ | $55.75 \pm 1.09$ |

"Taken as a whole, I think it will be found to answer the question, for the constants are uniformly the same in showing no effect of selection", GOWEN (48) states. But if I may express my opinion with regard to his results, I would remark that table 1 nevertheless gives the impression that this culture was not so indifferent as GOWEN thinks, for contrary to the intention of the experiment namely selection in the direction of low crossover, one notices a gradual rise in the c.o.v., which may be explained as due to inbreeding with the possible result that the flies had become homozygous for factors not taken into account, for evident reasons. Another possible explanation may be based on results of experiments done by other investigators, who found differences in crossing-over in connection with homogenous and heterogenous homologous chromosomes. One may imagine a gradual balancing of the two homologous chromosomes, which from the start might have been, genetically speaking, very 'asymetrical' with a consequent low crossover potentiality. Curiously enough GOWEN makes no further comment on this fact.

As regards selection towards high c.o.v.'s we notice only chance variations. GOWEN consequently was of opinion that his material was homogenous, without any heterozygous crossover modifiers, no matter in what direction selection was practiced. "The crossover mechanism is then working in a system of events controlled only by the mechanism used in crossing-over for the particular set of factors". These results together with GOWEN's conclusion cannot be generalised, for several later experiments have pointed out (a) the possibility of selection on c.o.v.'s; (b) that the difference in constitution affects crossing-over.

Apart from crossover-modifiers as such, several cases have been discovered where ordinary genes, acting pleiotropically, also affect crossing-over. We have in mind "The influence of the purple gene on the crossing-over between black and cinnabar" in the central region of the second chromosome of *Drosophila melanogaster*. This investigation was carried out by SEREBROVSKY (100) and was reported in 1927. Holding the opinion that the 'gene' constitutes a physical part of the chromosome, and advocating the presence-absence theory, SERE-BROVSKY considered the possibility of measuring the length of the gene, and especially in his experiments the purple gene (pr). His method was to compare the c.o.v.'s between black and cinnabar (b, cn.), obtained from the results of different experiments, when the chromosome was of different allelomorphic constitution each time, that is to say he compared c.o.v.'s obtained from the heterozygous female, now homozygous recessive for purple, and then heterozygous for it, and in still another case homozygous wild for this gene, e.g. $\frac{p}{p}, \frac{P}{p}, \frac{P}{P}$, resp. The difference in crossing-over between symmetrical and asymmetrical chromosomes would represent the length of the gene.

Another principle considered by SEREBROVSKY is the assumption that crossing-over is an absolute function of distance. For only in this case would there be possible a measurable asymmetry between heterogenous homologous chromosomes. He found that the value of crossing-over between b and cn turned out to be the greatest in the typical mating when between b and cn are substituted the normal allelomorphs for pr, i.e. PP (pr = p). The substitution of pp for PP diminishes the crossing-over, that is to say "shortened" the chromosome between b and cn. In the case of the asymmetrical construction of the homologous chromosomes, that is to say with Pp, there possibly occurred the most marked decrease in crossing-over, but the difference between Pp and pp turned out uncertain.

If recessive mutation means shortening of the chromosome then it will not be evident that a constitution $\frac{\text{wild}}{\text{b pr cn}}$ and a constitution $\frac{+ + \text{cn}}{\text{b pr} +}$ should give equal c.o.v.'s because of the consequent difference in degree of asymetry between the homologous chromosomes. The argument may be extended in connection with the total length of the chromosomes. With the accumulation of all recessives in one chromosome and of all normal allelomorphs in the other, it is clear that the asymetry will be greater than in the case of a proportionally equal distribution of recessive and wild genes over the homologous chromosomes in question. It is perhaps more illustrative to give a summary of the general total c.o.v.'s obtained for the different constitutions of the heterozygous females. The totals were as in the following table where A means eggs laid during the first six days and B eggs laid during the second period of six days.

TABLE 2.  The amount of crossing-over in females of different constitution which all emerged on the zero day. After SEREBROVSKY (100).

| constitution of females | value of crossing-over | | | |
|---|---|---|---|---|
| | between b and cn | | between Td and b | |
| | A | B | A | B |
| PP | 7.21 ± 0.29 | 3.68 ± 0.22 | 30.99 ± 0.58 | 27.84 ± 0.74 |
| pp | 5.36 ± 0.17 | 3.08 ± 0.22 | 28.94 ± 0.48 | 23.41 ± 0.63 |
| Pp | 4.53 ± 0.12 | 2.80 ± 0.15 | 28.74 | 27.10 |
| Ppcc | 5.01 ± 0.28 | 2.70 ± 0.24 | 32.37 | 25.91 |

From this table it appears that the constitution PP in group A as well as in group B shows the highest c.o.v. between black and cinnabar (7.2 and 3.7 resp.). The constitution pp is accompanied by a reduction in the c.o.v. to 5.4 and 3.1 for group A and B resp. "It seems as if a kind of shortening of the chromosome between black and cinnabar takes place here". The substitution of the heterozygous for the homozygous type e.g. Pp for PP reduces the c.o.v. to 4.5 and 2.8 for A and for B with regard to the region black-cinnabar.

It seems to me that according to these findings of SEREBROVSKY the c.o.v. obtained from a repulsion back-cross need not be necessarily equal to that of a coupling back-cross. In the latter case there will be a greater asymetry between the homologous chromosomes than in the case of a repulsion experiment where the normal allelomorphs and the mutant genes are more or less equally distributed over both homologous chromosomes, e.g. $\dfrac{AB}{ab}$ and $\dfrac{Ab}{aB}$.

Allelomorphic genes, therefore, distributed differently over the homologous chromosomes, may give different crossing-over values.

By reading SEREBROVSKY's article one comes under the impression that the writer perhaps attached too literal a meaning to the c.o.v. as a function of distance. It is particularly noticeable in his sentence just quoted. Nevertheless his experiments offer a working hypothesis which on itself is sufficient to justify research in this direction, and the facts obtained by him retain their great value. It was therefore a welcome fact when in 1929 SEREBROVSKY, IVANOVA and FERRY (101) published their results of further investigation in this direction in a paper: "On the influence of the genes y, li and Ni on crossing-over close to their loci in the sex-chromosome of *Drosophila melanogaster*" (y: yellow, li: lethal, Ni: Notch). Both latter mutations are lethal when present in homozygous condition. The method of investigation was essentially the same as in the former experiments on chromosome II. Changes in the c.o.v. were studied in connection with the substitution of different allelomorphs.

The results obtained showed that the lowest c.o.v. was found in females of that constitution, which in terms of presence-absence represents the greatest asymetry between the homologous sex-chromosomes.

The rise in c.o.v. obtained with regard to other cases, as compared with the decrease in crossing-over in the former may be looked upon as cases of smaller asymetry between the sex-chromosomes resulting from the absence of the normal allelomorphs for y, li, and Ni, respectively.

While geneticists in America and especially MORGAN's school, were very busy constructing chromosome maps or rather plotting the loci of the genes, the problem as to the constancy of the c.o.v. was of course duly considered. But workers who were not so intimately connected with MORGAN's school got the impression that this school assumes the c.o.v. as an absolute function of distance. The numerous cases proving the variability of the c.o.v. made others look with suspicion at the chromosome maps. Names that may be specially mentioned in this connection are DETLEFSEN, RASMUSSON and ROBERTS. They too were specially concerned with genetical factors causing changes in the linkage values. DETLEFSEN and ROBERTS (32) for instance performed successful selection experiments on c.o.v. in low direction, the region in question being white-miniature. The two first selections for series A and A' showed no or very little effect. Although these first selection values were few in number nevertheless a regression with reference to parental mean c.o.v.'s was apparent. Here the co-operation of genetical and environmental factors was strikingly evident, causing at the same time great difficulty in selection, for selection was only successful when the c.o.v. of the selected pair was genetically caused and not due to chance deviation from the M c.o.v. Nevertheless in the $F_3$ (low selection experiment) a female was selected which gave a c.o.v. of 17.99. Her offspring however, gave a mean c.o.v. of 26.18. But after the $F_5$ the procedure was enhanced. The $F_9$ gave 16.49 while the $F_{10}$ . . . . . $F_{13}$ eventually resulted in a c.o.v. of 0 %. This was serie A. A series A' derived from A would contain the proof. It was contained for 9 generations when at the 9th generation the M c.o.v. was 2.02 %. The $F_{14}$ gave a M c.o.v. of 0.44 and among this generation there were 25 pairs giving in total a M. c.o.v. of 0.2 while the mass inbreeding gave 0.63 % of crossing-over. The other stock maintained its value of 0 % for 9 generations. Reflecting that the original value was 33 %, undoubtedly one is convinced of the success of this low selection experiment.

Illustrative was the following case of series B which went through the process as follows: $F_1$ 28.6 %; $F_8$ 24 %; $F_{13}$ 10.2 %; $F_{29}$ 6.33 . . . . . $F_{50}$ 6.98 %, a remarkable fact as well.

It will be remembered that GOWEN's (48) high selection experiment gave negative results, which however might have been due to homogeneity of the factors concerned. DETLEFSEN and ROBERTS, too, obtained negative results for their high selection experiment, for to the contrary, the c.o.v. was lowered. DETLEFSEN and ROBERTS (32) are of opinion that GOWEN's negative results may be due to his experimental procedure. For he himself stated that his great difficulty in selection was the small number of individuals that could possibly be included in the 4th generation. Besides it is very difficult to judge the degree of selectivity exercised by him, for he only gave the mean total c.o.v. for each generation, and does not state the number of pairs he selected, neither does he give us a curve of distribution of the c.o.v.'s of such pairs, which was always done by DETLEFSEN and ROBERTS.

I think that the soundest argument in connection with the contrary

results of GOWEN on the one hand and of DETLEFSEN and ROBERTS
on the other is that GOWEN worked with factors in the 3rd chromosome
while DETLEFSEN worked with factors in the sex-chromosome.

One naturally wonders what actually did happen in connection with
the experiments done by DETLEFSEN and ROBERTS. The presence-
absence theory would lead one to think that, owing to deficiency,
white and miniature were brought so close together as to result in
complete linkage. Cytologically one asks the question whether the
frequency of twisting of the chromosomes as such had been reduced
between these two genes; one may also wonder whether not small
mutations of crossing-over modifiers occurred which in their turn
made selection in crossing-over possible.

## § 2.  *Drosophila simulans*

While DETLEFSEN and ROBERTS covered a rather large part of the
X-chromosome with only 2 factors, STURTEVANT's investigations
(113, 114) this time on *Drosophila simulans*, covered the greater part
of the sex-chromosome, involving several genes. The sex-chromosomes
of *D. melanogaster* and of *D. simulans* are very nearly the same, at
least they are far more similar than the V-shaped autosomes of these
two species, in which latter case weak corresponding of loci is a general
occurrence. General experience shows that crossing-over in *simulans*
is relatively constant.

A crossover reducer is located in the left end of the X-chromosome
as well as a crossing-over modifier in the right end of the 3rd chromosome. It is not certain whether reduction of crossing-over to the left
of the 2nd chromosome of *D. simulans*, obtained by crossing a multiple
recessive stock to a wild stock collected at New Orleans 1926, is due
to a crossing-over modifier (CIIL) present in the wild stock, or whether
it is due to an inverted section of the chromosome. Much more sure
is the following case reported by STURTEVANT (117). A stock of yellow-
prune-dusky-forked was crossed with various wild stocks. All the $F_1$
females gave a c.o.v. for yellow-prune that was much lower than the
standard value of 2.9 %. Also a reduction in the c.o.v. was obtained
for prune-dusky in which case a c.o.v. of 31.5 % was obtained as compared with the standard value of 39.3 %.

These experimental data convinced STURTEVANT of the presence of
a dominant crossover modifier in the y-pn-dy-f stock, which may

perhaps also be located in the X-chromosome. The corresponding values for the *melanogaster* X-chromosome are approximately 1.0, 37 and 20. It will be seen that the reduced y-pn value of *simulans* is much like the *melanogaster* value, but for the other 2 intervals the standard values are closer to the *melanogaster* values.

## § 3.  *Pisum*

These genetical changes in the c.o.v. do not pertain to *Drosophila* alone. They were also reported for *Pisum*, maize, *Gammarus chevreuxi*, and the Japanese silkworm, while Fräulein Dr. HERTWIG informed us personally that the c.o.v. for certain factors in the X-chromosome of the fowl is extremely variable. And it is possible that genetic causes may be traced later on.

RASMUSSON (93) gave us a survey of the situation with reference to *Pisum*. He pointed out that MENDEL had found independent assortment of the genes Le (length) and V (pod parchment). RASMUSSON however, found that in his case these factors were linked. Further investigation involving the factors Bta (blunt pod) and P (semi-parchmented) brought to light the same phenomenon of the variability in crossing-over. The linkage group V-Le-Bta-P was found, but a great variability in the c.o.v.'s for the different regions was observed. These variations might be due for the most part to genetic causes according to RASMUSSON. *Pisum* research in this direction is urgent. We are reminded of HAMMARLUND's results, which showed a strong linkage for certain genes among some of the $F_2$ offspring, while others showed independent assortment. This was found in connection with other experiments as well. Absence of linkage in an $F_1$ mated to the recessive parent plant suggested that the cause of the linkage was not inherited independently of the two linked factors, because if it were, then we should expect at least some of the crosses to show linkage.

RASMUSSON examined the linkage group Btb-Cp (pod apex and pod form) and observed a statistically significant difference in linkage among the different crosses.

HAMMERLUND (88) studied the linkage between the genes A and Gp (flower colour and yellow pod) and found very different linkage values. A purple flower (A) — green pod (Gj) was crossed with another variety with white flowers and yellow pods. The $F_2$ of several $F_1$

plants showed strong deviating results, some showed independent assortment while others showed a very strong linkage.

SVERDRUP (120) found differences in crossing-over with regard to crosses between several types of *Pisum*. She also found differences in crossing-over between male and female gametes. It concerned the factors K and W (normal wings of the flower and leaf blossom); B and St (flower colour and normal stipulae).

WELLENSIEK (128, 129, 130) found that the factors P (pod colour), Cp (pod-apex) and V (pod parchment) are linked in the one crossing and not in another. He furthermore found that weakening in linkage between two factors caused a weakening in linkage among factors belonging to the same group. The identity of the factors in *Pisum* work however remains an open question.

KAPPERT (93) found linkage between the factors R (cotyledone form) and Bt (pod apex); RASMUSSON however found no such linkage. But according to KAPPERT his cross concerned the varieties *Pisum Thebaicum* and ,,*Wunder von Amerika*''. RASMUSSON is of opinion that it might be a cross between *Pisum Thebaicum* and the variety *Witham wonder*, in which case there would be a great probability as to the identity of factors R, and Bt of KAPPERT and R and Btb or perhaps Bta of RASMUSSON. Consequently the difference in results may be due to genetic causes. Very illustrative in this connection is the summary given by RASMUSSON with reference to the variability of crossing-over in *Pisum* (93, p. 116).

One comes to the conclusion that the disharmony in cytological and genetic results in *Pisum* work as it stands today may be nothing more nor less than a result of genetic changes in the c.o.v. and of their non-identity of the genes concerned in the different varieties worked on. Consequently, no matter how diligently the *Pisum* work is continued this disharmony will remain in its cyto-genetics. It is therefore a very urgent and important matter for *Pisum* workers to come to some general agreement as to their line of action. Among others it should be agreed to start afresh with homogenous material that is to say varieties with their corresponding loci identical. This does not mean that work in the direction thus far followed must be stopped, for it has given much insight and should be continued but this time with the full appreciation of the possible non-identity of the corresponding loci among the different varieties worked on.

## § 4. *Zea mays*

We may now turn to *Zea mays* the genetics of which has attained a level second to that of *Drosophila*.

STADLER (102) found that the c.o.v. for the regions C—Sh, Sh—Wx, C—Wx differ significantly among different families for male as well as female gametes. C = aleurone colour; Sh = shrinkage of endosperm Wx = chemical composition of endosperm.

The most striking difference was noted in connection with plant 4017-5 and 4013-2, which happened to be planted on the same day and grew up under similar environmental conditions within 5 yards from each other. No. 4017-5 showed a striking difference with 4013-2 for the regions Sh-Wx, C-Wx, of $11.3 \pm 0.9$ and $10.2 \pm 0.9$ resp., the difference being approximately 12 and 11 m. diff. respectively.

COLLINS (25) and KEMPTON also studied the variability of linkage values for the two seed characters C and Wx in maize. It was observed that some progenies were far more uniform than others as regards the stability of crossing-over between C and Wx. More extensive study of some progenies for several generations displayed a statistically significant difference in crossing-over. Of particular interest was the fact that when the plants were divided into two classes, the one homozygous and the other heterozygous for R(aleurone), then crossing-over was much lower for the latter class. This was analogous to SEREBROVSKY's results in connection with the influence of the purple gene on crossing-over in the region black-cinnabar, discussed above.

The same phenomenon was observed by COLLINS and KEMPTON with regard to the gene Su (sweet endosperm), that is to say the constitution Su su effected a smaller c.o.v. than the constitution Su Su. Furthermore it was observed that the reciprocal crosses differed. In some progenies the c.o.v. was greater among male gametes, in others it was greater among the female gametes. Seeds at the apex of the cob showed a greater c.o.v. than seeds at the base, but this was the case only for female gametes.

## § 5.  *Gammarus chevreuxi*

HUXLEY (60) compared the c.o.v.'s of different stocks of *Gammarus chevreuxi* and obtained striking evidence that variations in crossing-over in the male was due to genetical causes. From his table 6(60, p. 151—153) it is apparent that the totals for the stocks CPLC26 and CPLC45 and Rep. y$\alpha$, gave 37.6; 21.4 and 18.9 % crossing-over between the factors b and c. Stock CPLC45 showed something extra-ordinary, for the male had a very low c.o.v. of 9.3 % but his offspring gave a mean c.o.v. of 31.4 %.

## § 6.  *Silkworm*

In the silkworm KOGURA MAKITA (69) studied the ,,Modifying factors influencing the striped-yellow crossing-over value''. In the silkworm no crossing-over was observed in the females. Considerable variation in the percentage of crossing-over was observed for the factors S (striped larval marking) and Y (yellow blood colour), and this led STURTEVANT to suggest the existence of a modifier. MAKITA succeeded by selection to obtain families with a high and families with a low c.o.v. The families with a high value maintained this high value and segregated no low value families. Low families segregated intermediate and higher values. Individuals with an intermediate value produced offspring of high, intermediate and low values in a ratio more or less 1 : 2 : 1. This suggests the presence of a dominant modifying factor CII producing the c.o.v. between S and Y.

Although I am sure that these cases briefly surveyed by me do not exhaust the facts so far obtained in this connection, yet I think that hereby the great importance of genetical causes in connection with the changes in the linkage values has been sufficiently emphasized.

# CHAPTER IV

## The effect of age on crossing-over

### § 1. *Drosophila melanogaster*

It is very difficult in the case of temperature effect on crossing-over to make out whether the germcells are influenced directly or whether it is due to the general physiological condition of the animal or plant treated. But I think we may safely say that in the case of age influence the effect on crossing-over is due to the general physiological condition of the animal or plant concerned.

Several experiments on *Drosophila* have shown the effects of age on crossing-over. The method usually followed was to compare the c.o.v. calculated from the first brood with that of later broods, other things being equal.

The results published by BRIDGES (17) show the influence of age on crossing-over in the second chromosome of *Drosophila melanogaster*. As to the X-chromosome, no age influence on crossing-over could be detected. This is perhaps a striking argument in favour of the physiological individuality of the linkage groups. As regards the second chromosome BRIDGES (19) back-crossed a $F_1$ female derived from the coupling cross wild × purple-vestigial. In 8 cases the c.o.v. of the first brood was compared with that of the second brood and for seven of these cases a decided decrease in crossing-over was observed for the second broods. In order to fulfil the requirements of balanced inviability, also a repulsion cross was done the results of which showed the same phenomenon. Of the six heterozygous females all gave relatively lower c.o.v.'s for the second broods.

Also for the region purple-curved this age effect on crossing-over was observed.

Some of the earlier data however, suggested that this deviation from the standard values was recovered after the 2nd brood, perhaps even surpassing the standard values. Consequently experiments were planned to study the effect of age during the lifetime of the fly. The c.o.v.'s were compared untill the 4th brood. Very interesting results were ascertained. The loci involved this time covered nearly the entire length of the 2nd chromosome, but they were comparatively few in number, being as follows: $\dfrac{S + + +}{+ \; pr \; c \; sp}$, S = star; pr = purple; c = cur-

ved; sp = speck. A heterozygous female back-crossed to a recessive male gave normal c.o.v.'s for the first brood, the expected decrease for the second and a small further decrease for the third and the fourth broods. Unfortunately however, this experiment was at a disadvantage because of the long regions which allowed for double crossing-over.

Furthermore, it will be evident that 10-day classes may obscure facts which otherwise would have displayed themselves, for instance in the case of shorter class intervals. As stated elsewhere in this chapter, PLOUGH was able to overcome the latter objection by making use of 2-day classes, which proved to be of very great importance, for PLOUGH hereafter actually found, that, although there was some variation, yet a typical rythmical fall and rise were apparent in the c.o.v.'s, "which must be the expression of fundamental physiological changes in the development of the female"; BRIDGES (17). As regards crossing-over between black and purple the curve behaved as follows. The first values were relatively high, 8 %, which decreased quickly for the 9 subsequent days and then slacking down again, untill the value of 5 % was reached and maintained with very little change untill the 21st day. The next decrease was again slow reaching a value of 3.5 % on the 30th day. Hereafter it showed a steady rise. But unfortunately at the 25th day the numbers were too small for any decisive judgement. It was suggested that the fall in the c.o.v. must be brought in relation to co-incidence in so far as the internodal region in some way or other was changed in regard to its length.

PLOUGH's (89, 90) experiments showed still more. BRIDGES did not completely study the left half of the 2nd chromosome. By making use of the mutation Star, PLOUGH was able to make a further study of the age effect on crossing-over in this region of the 2nd chromosome. It was found however that the greatest change in the c.o.v. took place in the central region of this chromosome, viz. the regions black-purple-curved-vestigial. The regions to the left and to the right were less subject to changes in crossing-over.

PLOUGH (90) however continued his experiments and in 1921 he reported the results of his continued studies. His former results (1917) as well as those of BRIDGES (1915) were fully affirmed namely the absence of age effects on crossing-over in the X-chromosome. (However as we will see furtheron such an effect was indeed observed, but at the very extreme right end of the chromosome which end corresponds to the place of attachment of the spindle-fibre.) As regards the 3rd chromosome, it was observed that the sections from sepia to Dichaete and from Dichaete to spineless were effected to a lesser degree by age, while temperature had a far greater influence. For the sections spineless to sooty and from sooty to rough neither age nor temperature was observed to exert any notable influence on crossing-over.

The absence of changes in crossing-over due to age influences in the X-chromosome as well as a possible mistake in regard to a small rise in crossing-over obtained for the 3rd chromosome for the 2nd broods, instigated BRIDGES (17) to continue his experiments and in 1927 he reported his results ascertained in connection with the relations of the age of the female to crossing-over in the 3rd chromosome.

The results of two very important experiments were reported, namely those with age classes of two days, as PLOUGH had them, and those results which showed genetic identity in the c.o.v. for 8 different cultures. The totals could therefore be used for the construction of curves relating to age effect on each region respectively, the graphic representation of which may be seen in Fig. 4, (17, p. 76).

Inspection of these curves shows a commencing high recombination percentage, viz. 71.7 % which value however, suddenly drops to a minimum of 45.7 % at about the 11th day; a second maximum is attained on the 19th day. The following sudden fall and recovery is again followed by a small gradual rise, so that the rythmic phenomenon suggested by PLOUGH's (89) experiment was fully affirmed. By studying the respective recombination curves however, a difference in effect will be observed for the different regions. With regard to the difference in decrease (expressed %) of amount of recombination between the first high value and the first minimum, the order was as stated in table 3, p. 36.

Again, the greatest difference (91 %) was observed for the region st-pp, that is to say for the midsection of chromosome III. The sections immediately to the right and immediately to the left of the midsection i.e. pp-ss and D-st, were also strongly affected to more or less the same degree, but not to such an extent as the midsection,

TABLE 3. Age effect on crossing-over in different regions of chromosome III. After BRIDGES (17).

| section | second experiment | second experiment | first experiment | average decrease % |
|---|---|---|---|---|
| | | decrease % | decrease % | |
| st-pp . . . | 5.6 to 0.5 = 5.1 | 91 | 81 | 86 |
| pp-ss . . . | 13.4 to 4.2 = 9.2 | 69 | 62 | 66 |
| D-st . . . | 1.7 to 0.6 = 0.0 | 65 | 52 | 59 |
| Total | 71.7 to 45.7 = 26.0 | 36 | 36 | 36 |
| h-D . . . . | 12.7 to 8.4 = 4.3 | 34 | . . | 34 |
| ss-es . . . . | 12.8 to 8.8 = 4.0 | 31 | 37 | 34 |
| ru-D. . . . | 36.1 to 29.8 = 6.3 | 17 | 21 | 19 |
| ru-H . . . | 25.6 to 21.2 = 4.4 | 17 | . . | 17 |

the percentages being 69 and 65 respectively. The 2nd sections on either side of the midsection were similarly affected but only to half the extent of the former sections, the differences being 31 to 34 %. Lastly, the section furthest removed from the midsection e.g. ru-H, showed the smallest difference, 17 %.

It reminds of a binomial distribution but of course it has nothing more in common with it. It simply corresponds with the peculiar shape of the chromosome in question. In connection with this I may refer to the study of ANDERSON and RHOADES (Papers of Mich. Acad. of Sc. Arts & Let. 13, 1930 on "The distribution of interference in the X-chromosome of *Drosophila*." They pointed out that the interference for a given map distance is not equal for the different regions of the same chromosome. The most marked differences were noted in the 2nd and 3rd chromosomes where interference is lowest around the midpoints and much higher towards both ends. "The symmetrical distribution about the midpoints of chromosomes II and III is probably related to the fact that these chromosomes are V-shaped and that their spindle-fibre is attached at the midpoints". In the X-chromosome smaller differences were observed but the data showed clearly that interference is greater in the left half of the chromosome than in regions further to the right. They worked on the material of BRIDGES and OLBRYCHT 1926 and of ANDERSON 1925, and manipulating Haldane's formula: $y = 0.5 (1-e^{-2x})$, where y is the amount of recombination, $x$ the amount of crossing-over and e the base of the natural logarithms, they were able to study more accurately the relation of recombination to crossing-over, between interference and map distance. They found interference lowest at the right or spindle-fibre end of the X-chromosome, increasing gradually to the left or distal end.

It being known that the middle sections of the 2nd and the 3rd chromosome are specially liable to age influences on crossing-over BERGNER (10) in 1928 studied the effect of prolongation of each stage of the lifecycle on crossing-over in the foresaid chromosomes of *Drosophila melanogaster*. The heterozygous $F_1$ females of the consti-tution II $\dfrac{b\ pr}{+\ +}$ III $\dfrac{+cu}{D\ +}$ were back-crossed to b-pr-cu males (II, III = 2nd, 3rd chromosome; cu = curled). BERGNER made use of controls to obtain standard values for ordinary age effect with the typical rythmic W-shaped curve.

Experiments on the prolongation of the stages in the lifecycle, were carried out on $F_1$ females that upon emergence were kept from laying eggs for several days as a result of narcotic treatment and of low diet. At the end of the treatment of temporary enforced sterility they were crossed and further treated thereafter in the same manner as the controls. Also the pupae and the larvae were treated for the prolon-gation of the stages in question, the pupae being nearly deprived of Oxygen and the Carbon dioxide kept from escaping; the larvae were held on low diet.

The result was that prolongation of different stages in the lifecycle actually had an influence on the c.o.v. In the 1st instance a relatively sooner inset of decrease of the c.o.v. was obtained for early prolon-gation of the egglaying cyclus. This first drop lasted for a shorter period than was found in former experiments, and the second rise appeared sooner. Failing of changes in the c.o.v. in cases where the flies which emerged first were kept from laying eggs for four or five days, suggest that the eggs in oöcyte had passed the socalled sensitive period when age exerts its influence on crossing-over. Similarly the time of disappearance of the effect in cases of prolongation of the larval stage suggested that the early oögenesis was not affected. In this manner BERGNER came to the conclusion that the early middle stages of oögenesis seem to be the most liable to changes in crossing-over due to age effects.

Concluding the data on *Drosophila melanogaster* we have seen that one of the most important results obtained is that of the relatively high sensitivity, also to age influences, of the middle region of both the 2nd and the 3rd chromosome, these regions corresponding with the place of attachment of the spindle-fibre. This conjecture was fully substantiated genetically and cytologically. Cytologically, ANDERSON (4) found that the place of attachment of the spindle-fibre to the X-chromosome is at the right end of the chromosome. STERN (105) thereupon studied the age influence on crossing-over between genes located in the immediate neighbourhood of the place of attachment of the spindle-fibre, namely the genes Bar and bobbed. He actually observed a small but significant drop in the c.o.v., which attained its minimum on the 4th day broods, whereafter it gradually recovered again. The 1st minimum was smaller in change and was sooner apparent than was found to be the case for the 2nd and the 3rd chromosome. These are very important results.

By means of studying the age effects on crossing-over, we came to ascribe special significance to the place of attachment of the spindle-fibre and consequently our customary genetic horizon is broadened. But it will now be apparent to the reader, that the warning of DETLEFSEN (29) that the crossing-over value is not to be taken literally as a function of distance must not be emphasized too much so as to give the impression that the c.o.v. is absolutely devoid of such a function. For genetically the central map regions do correspond to the central region of the V-shaped chromosomes and the right end of the genetical chromosome does correspond with the end of the cytological chromosome.

## § 2.  *Drosophila simulans*

It would be interesting to know what data are on hand in connection with the age effect on crossing-over in *Drosophila simulans*.

STURTEVANT (117) reported experiments in connection with the 1st and the 2nd chromosome of this species. The regions involved were sd-pm (sd = spread; pm = plum) at 80 and 103 of the *simulans* map. They cover the rightmost quarter of the 2nd chromosome. No age effect was observed. As regards the X-chromosome he studied the age effect on crossing-over between ruby (rb), dusky (dy) and forked (f), which loci roughly cover the greater part of the sex-chromosome. For the three experiments carried out a small rise in the c.o.v. was noted during the first 7—8 days, whereafter a decrease set in. However, none of the corresponding regions in *D. melanogaster* showed any age effect on crossing-over. Unfortunately the experiments were not directed to test the middle regions of the 2nd chromosome nor to test the right end of the X-chromosome.

### § 3. *Vertebrates*

Naturally it would be interesting to know what the age influence is on crossing-over in mammals e.g. rats and mice. No such age effect was observed by CASTLE and WACHTER (22), at least they think that their results cannot be interpreted as such. But the clearest case so far recorded for *Vertebrates* and which has direct bearing on the problem, is that reported by HALDANE and CREW (56) in connection with poultry. They definitely found a weakening of linkage between the factors B and S (barred feathers and silver hackles) with increasing age of the male fowl. The c.o.v. was found to be 21.9 $\pm$ 1.4 for the 1st incubation year, 36.9 $\pm$ 2.9 for the 2nd and 47.6 $\pm$ 3.6 for the 3rd, when nearly all linkage has disappeared. The authors remarked: "The changes observed by us might be due to diminished rigidity of the chromosomes, increase of the forces tending to break them or other causes". Apparently the phenomenon of rythmical rise and fall of the c.o.v. observed in the case of *Drosophila melanogaster* was absent or could not be detected.

### § 4. *Zea mays*

As far as plants are concerned, we may remind the reader of STADLER's (102) experiments on maize, discussed elsewhere in this treatise. A small but significant negative correlation was observed between crossing-over and the date of 1st pollen, for the region Sh-Wx. The region C-Sh remained indifferent. STADLER also studied individual cases in the male gametes, but observed no definite influence. In order to do this, pollen was taken from 4 heterozygous plants every day or two and homozygous plants pollinated, which process was carried on throughout the whole pollen shedding season.

STADLER's (102) data concerning female gametes were less certain. However, crossing-over was found to be regularly higher in tiller ears than in ears on the main stalk. Only in 4 out of the 21 cases a higher c.o.v. was noted for ears on the main stalk, but only one of these was statistically significant. It is very difficult however, in this case to discriminate between age effect or other milieu influences on the c.o.v., but if age was responsible for the change, then the experiments point out that the c.o.v. increased with age, for tiller ears are formed later than ears on the main stalk. To me it is not clear that a tiller ear should be considered younger compared with ears on the main stalk, for a tiller is a seperate plant, which as such does not allow for direct comparison, or at least its dependence on the main stalk may not be always the same.

It is a pity that so little has been done on plants in this direction. I therefore commenced studies on *Pisum*, making use of the factors acacia leaf and wrinkled seedcoat. The c.o.v. is to be calculated on the $F_2$, the intention being to compare the values obtained for the earlier till the later pods, making use of 2-day classes for example.

Summarising, age has a typical effect on crossing-over in the central regions of chromosome II and III and a slight but similar effect on crossing-over in the right end of the 1st chromosome of *Drosophila melanogaster*, the latter region as well as the central regions of chromosome II and III being the place of attachment of the spindle-fibre. Detailed study showed that the age effect on crossing-over may be expressed in a W-shaped curve agreeing with a first decrease in the c.o.v. then a rise, a second fall etc. No age effect was observed for the second chromosome of *Drosophila simulans*. For the first chromosome there was a rise for the first 7—8 days whereupon a decrease set in. None of these corresponding loci in *Drosophila melanogaster* showed age effect on crossing-over. No age effect was detected for crossing-over in rats and mice. A sure positive correlation was observed between age and crossing-over in poultry. In *Zea mays*, female gametes on tiller ears showed a higher c.o.v. than female gametes of ears on the main stalk. A small but negative correlation was observed between crossing-over and date of first pollen for the region Sh-Wx, while the region C-Sh remained indifferent.

## CHAPTER V

### THE EFFECT OF SEX ON CROSSING-OVER

§ 1. *Zea mays*

One of the best pieces of work on the variability of crossing-over among male and female gametes is that of L. J. STADLER (102) on *Zea mays*. The chromosome regions which he studied were C-Sh-Wx (C affects aleurone colour in the presence of certain other genes; Sh affects the degree of shrinkage of the endosperm at maturity; Wx affects the chemical composition of the endosperm). The c.o.v.'s as extensively studied by KEMPTON (1919) and HUTCHINSON (1922) were 3.5 for the region C-Sh and 21.88 for the region Sh-Wx.

In order to avoid complications in connection with the complementary factors of C in the production of aleurone colour, it was seen to that the parents were homozygous for these genes to wit: A, R and i.

The results obtained were very interesting and for the physiology of crossing-over of very great importance. The variability of the c.o.v. was of the same order for both male and female gametes to wit: the coefficients of variation were greater for the shorter region C-Sh than for the longer region Sh-Wx. But for each region the female gametes showed a slightly higher coefficient of variation than the male gametes. This variability was statistically proved to be well beyond that which might be accounted for by fluctuations of sampling.

While in the case of male gametes no definite correlation was observed between the c.o.v. and the characters indicating vigor of growth, in the case of female gametes however there seemed to be a clear tendency towards negative correlation between crossing-over and these characters. Date of first pollen shedding showed no significant correlation with the c.o.v. in the case of female gametes, while for male gametes there was a slight but possibly significant negative correlation, but only for the region Sh-Wx. It was also found that the c.o.v.'s for female gametes were significantly higher in tiller ears than in ears on the main stalk, which phenomenon was not observed in the case of male gametes. Possibly this might be due to age influence. As far as differences between male and female gametes of the same plant were concerned, the c.o.v. was definitely found higher in the case of male than in that of female gametes. The difference ranged from 8 to 12 x P. E. diff., so that the statistical difference could not be questioned.

At this stage we are almost convinced that this difference in c.o.v. was associated with sex, yet STADLER is inclined to think that it was not. The cases reported by EMERSON and HUTCHINSON (1921), in which the c.o.v. in the region C-Sh was higher in the female gametes, together with other trials in which no significant difference was found, supported the view that it was not primarily a question of sex difference but one of age. This may be the case when one considers ears on the main stalk older than ears on tillers of the same plant, for such differences have been found. STADLER argued further that female gametes developing at different times, show different c.o.v.'s, and it is therefore possible that this age factor was the primary cause of the difference in the crossing-over between male and female gametes, for microsporogenesis and macrosporogenesis also occur at different periods. But I do not agree with this argument, nor am I convinced that EMERSON and HUTCHINSON's (38) conclusion support STADLER's view of the matter to any great extent. Their conclusions, though to a lesser degree, are subject to the same mistake as that of DETLEFSEN (30) elsewhere mentioned in this chapter, who found for each individual case in question, no difference statistically significant enough as to conclude to a difference in the obtained values, yet closer examination would have shown that the deviations, however insignificant, were all in the same direction. Besides EMERSON and HUTCHINSON (38) themselves state that where $F_1$'s were used as pollen parents the observed distribution deviates so markedly from that expected on the basis of the same frequency of crossing-over in microspore as in megaspore development, 3.13 %, that the difference cannot be ascribed to chance. It would be seen therefore that there is a significant difference in the frequency of crossing-over in megaspore and microspore development so far as the C and Sh factors are concerned. Such a difference is further indicated by the fact that out of 35 $F_1$ plants tested 30 i.e. more than 85 % showed a greater frequency of crossing-over in megasporogenesis than in microsporogenesis, while in only 5 of the cases the reverse was true.

BREGGER (14) found the c.o.v. between aleurone colour and waxy endosperm 2.46 higher in male gametes. This agrees well with the difference of 2.9 found by STADLER.

In connection with B (plant colour) and L (liguleless) EMERSON and HUTCHINSON (38) found the c.o.v. of 19 $F_1$ plants for microsporogenesis 38.01 ± 0.52 and for megasporogenesis 36.58 ± 0.50, a difference of 1.43 ± 0.72 or 1.99 × P.E., which means that the difference is not statistically significant. However for 19 $F_1$ plants tested for individual differences it was apparent that for 8 plants the c.o.v. was higher in megasporogenesis, while for the remaining 11 the c.o.v. was higher in the case of microsporogenesis. It is quite possible that there may exist a hereditary basis for these individual differences between micro- and macrosporogenesis.

Again, EYSTER (39) found in connection with the linkage between tunicate ear (tu) and sugary endosperm (su) a higher c.o.v. for microsporogenesis, being 34.85 % approximately for male gametes and 26.94 for female gametes. Besides the c.o.v.'s for the latter showed a greater variability than those of the former. STADLER (102) also found a greater variability among the female gametes for the regions C-Sh, Sh-Wx and C-Wx.

BREGGER (14) also found the c.o.v. greater for male gametes by 7.91 for tunicate ear and sugary endosperm.

Our total impression therefore acquired from the data on maize discussed above is that sex does influence crossing-over, the percentage generally being higher for male than for female gametes.

## § 2. *Pisum*

Although comparatively easy, still experiments differentiating between crossing-over among male and female gametes in *Pisum* are very rarely met with. Undoubtedly it is of very great importance especially in connection with *Pisum*, for if the c.o.v. differs significantly between pollen and ovules then of course our average c.o.v. obtained from both may greatly mar our results and our insight in the linkage groups of *Pisum*. Fortunately however the data so far obtained show no significant difference between c.o.v. obtained from male and from female gametes of this plant. As regards the *Pisum* linkage group 2, MISS DE WINTON (139) states in her survey, that back-cross and reciprocal back-cross experiments were carried out in connection with the factors y-gr (yellow cotyledon, green pod.) By calculating the c.o.v. for pollen and ovules seperately, 4 % and 3.8 % were obtained respectively, thus almost the same values; when calculated on the basis of 3.9 % the deviation in both cases is insignificant.

As far as linkage group 3 is concerned crossing-over was studied in connection with purple flower — salmon flower (Bb); normal stipules — reduced stipules (Ss); purple pod — green pod (Vv). The c.o.v. for B-S + S-V = 37.7 % for female gametes and 38.6 % for male gametes, showing a rather strong interference for both male and female gametes. On the whole however the differences are so insignificant that we may conclude that crossing-over takes place with the same frequency for male and female gametes of *Pisum*.

§ 3.  *Primula sinensis and Pharbitis Nil.*

As far as *Primula sinensis* is concerned a rather troublesome case is met with, for it seems as if the c.o.v. is greater in the one case for male gametes and in the other for female gametes, notwithstanding the fact that it pertains to factors in the same chromosome. This is of great importance for logical induction. "In group I", states Miss DE WINTON (139), "two factors show 10 % of crossing-over with each other, the other two about 2 % with each other, but the two pairs are some distance apart. We have large numbers from back-crosses for testing these linkages and the first two do show definitely that the linkage is closer on the female side than on the male side whereas the two most closely linked factors show less crossing-over on the male side than on the female side".

With regard to the linkage group short style-long style (Ss); blue flower-not blue flower (Bb); green stigma-coloured stigma (Gg); light red leaf-dark red leaf (Ll), a great difference in crossing-over between male and female gametes was found, but the totals involved were relatively small, and the distances between the factors were of such dimensions that double and triple crossing-over was not excluded. However, this refers to both kinds of gametes. The percentage of crossing-over, except for G-L, was higher in male gametes, but I may remind the reader that GOWEN's (48) biometrical study disclosed a greater variability in crossing-over for shorter regions, so that the G-L case above mentioned might have been due to chance variation.

In 1916 ALTENBERG (6) studied the linkage between the factors Ll-Rr (magenta flower-red flower) and Ss. From his results it seems as if crossing-over between R and S takes place more often in macrosporogenesis, while for the region L-R crossing-over seems to take place more often in megasporogenesis. In the former case the c.o.v.'s were 32 for the male, and 27.2 for the female gametes, while in the latter case the c.o.v.'s were 0.9 for the male gametes and 2.8 for the female gametes. As far as double c.o. is concerned, we notice a small difference in favour of the male gametes. But I am not sure whether these differences can be wholly relied on for n = 392 only as regards the heterozygous female. Furthermore the difference in c.o.v. between male and female gametes for the 1st region is statistically insignificant. For the 2nd region the difference is perhaps significant, D/m being greater than 1, so that the general impression gained from ALTENBERG's data is that c.o. may be the same for male and female gametes in some cases, while in others c.o. may favour female gametes, or better still: the general conclusion arrived at from Miss DE WINTON's and from ALTENBERG's results is that c.o. may, in some regions of the chromosomes of *Primula* be equal among male and female gametes, in others greater among female gametes, while in still other regions it may be in favour of male gametes, so that our logical induction seems to lead to no apparent general rule as far as c.o. among male and female gametes in *Primula* is concerned.

IMAI (61) very cautiously states, that, although he found that c.o. takes place with a frequency probably the same for macro- and micro-sporogenesis in *Pharbitis Nil*, yet he would not state this as conclusive, for experiments, specially directed towards this problem have not yet been carried out by him.

## § 4.  *Drosophila*

With reference to insects, naturally, we first consider *Drosophila*. On the whole there is no c.o. in the male *Drosophila*, and this does not pertain to the X- and Y-chromosomes only but also to all other autosomes in the male. This is a remarkable phenomenon indeed. And yet very little theory is met with in connection with it. That c.o. does not occur between the X- and the Y-chromosomes is perhaps not so strange for morphological and perhaps also for physiological reasons. It is true that there are c.o. modifiers and we could have suspected the presence of such a modifier in the Y-chromosome, but for the fact that certain experiments have shown that c.o. occurs in the ordinary manner when the chromosome constitution of the female is XXY. (5). I cannot remember having come across experiments in connection with XO males, for generally they are sterile. If such experiments were possible it would have helped a great deal to solve the problem. It must be admitted that in the XXY case it is quite possible that the quantitative side of the question plays its part as well, for we have seen that in some cases the c.o. modifier may be overruled by other factors, its action being annihilated. Although it seems improbable that a factor present in the Y-chromosome would prevent itself from crossing-over, yet it is possible. If not, then surely it may cross over to the X-chromosome, thus imparting itself to females also, so that failing of c.o. would then be quite a commonplace phenomenon in the female as well, while c.o. would then also take place between X and Y. This generally does not happen in *Drosophila*, so that in the case of a c.o. modifier located in the Y-chromosome it must be of such a character that itself is also prevented from crossing-over, that is, it must be sex limited.

However it seems simpler to me to ascribe the whole phenomenon as due to a secondary enzyme action preventing c.o. in the chromosomes. It is to be hoped that aberation experiments of STERN and others will ultimately throw more light on the problem. The effect on c.o. of different pieces of the Y-chromosomes may be studied in cases of translocation etc.

Phenotypically several cases of crossing-over in the male *Drosophila* have been reported. But there are possible explanations, showing that these c.o. males have not resulted from the usual gametic crossing-over process. These alternatives are mutation, inversion, deficiency and duplication, wholly and partly somatic crossing-over. Granted that these cases are highly exceptional true crossovers, we may perhaps explain them as a result of failing of the male enzyme action on crossing-over in the male *Drosophila*.

### § 5. *Locust*

Apart from the order *Diptera* (*Drosophila*) 2 other land *Arthropoda* have been studied in this connection; in the order *Orthoptera*, the grasshopper, and in the order *Lepidoptera*, the silk worm; of the water insects we may mention *Gammarus chevreuxi*.

NABOURS (84) states that in the grouse locust, *Apotettix*, crossing-over is demonstrated as occurring in heterozygous individuals reproducing parthenogenetically as well and apparently to the same extent as in those females reproducing bisexually. Crossing-over occurs in the male as well but not to the same extent as in the female, that is to say, as far as the data went at that time. Perhaps the data were not sufficient to allow for a final conclusion.

For *Paratettix texanus*, NABOURS found the series of multiple alle-lomorphs A, B, C, D, E, F, H, I, J, L, N, P, Q, S, and a factor H, all producing different colour patterns. Closer investigation however showed linkage between A and H with a c.o.v. of 24 % in spermatoge-nesis, and 46 % in oögenesis, a much higher c.o.v. therefore for the female gametes. HALDANE's (55) experiments on this insect showed that 335 sperms from 1421 of diheterozygous males were crossovers giving a percentage of 23.6. The c.o.v., according to HALDANE's table 1 (55), vary significantly with different pairs of the allelomorphic series. Where A is involved 53 out of 180 were observed to be cross-overs, thus giving a c.o.v. of 29.4 %, while for the other factors to-gether the mean c.o.v. is 22.7 %. With the supposition that B, C, D are located at different loci in the same chromosome and that they act as c.o. modifiers, (crossover reducers) it should be expected that linkage would be stronger in a threefold heterozygous ♂ than in a diheterozygous ♂.

From his data it appears that in diheterozygous females crossing-over is 50 % while in triheterozygous females it is 45 %, n was large in both cases, 1570 and 1389 respectively. Consequently only offspring obtained from diheterozygous parents can be compared as to differen-ce in the c.o.v. with regard to sex. The same refers to triheterozygous parents. We then get the following comparison.

TABLE 4. Difference in crossing-over in ♂ and ♀ *Paratettix texanus*. From HALDANE's data.

| Diheterozygous parent | Triheterozygous parent |
|---|---|
| c.o.v. for ♂ gametes 29.4 | 22.7 |
| „  . . ♀  . . 50 | 45.4 |

We may conclude that for *Apotettix* and for *Paratettix texanus*, the data so far on hand show a decided higher c.o.v. for the female as compared with that of the male.

## § 6. *Gammarus chevreuxi*

HUXLEY (60) studied the sexual difference of linkage in the crusta-cean, *Gammarus chevreuxi*. The genes b (red-eye) and c (albino-eye) are linked. HUXLEY made coupling and repulsion experiments as follows: (a) BC . bc × bbcc, (b) Bc . bC × bbcc. Linkage was found to be stronger in the male than in the female, for the former a c.o.v. 25.4 and for the latter 50.6 % was obtained. There was a difference in viability among the classes, for the double recessives were less viable than the other. The c.o.v.'s with reference to repulsion experiments were higher than those calculated on data obtained from coupling experiments. HUXLEY also studied the variability of the c.o.v., con-structing a curve with classes of 5 % c.o. for male as well as for female gametes. The curve for the c.o.v.'s for female gametes is drawn out to the left with a modal value more than 50 % and a mean of 50.6 %. The modal c.o.v. in the male gametes varies between 20 and 25 %. The curve for the c.o.v.'s in females quickly accumulates to the right ending in classes 55—66 %. This of course agrees with the fact that 50 % represents the theoretical limit of crossing-over. HUXLEY is of opinion that it is possible that independent assortment takes place in the female *Gammarus*. This I think, offers great difficulties for the problem of localization, for then it would be possible for a gene in one chromosome to behave as an allelomorph towards one in another non-homologous chromosome.

It would have been possible that in connection with two attached non-homologous chromosomes, genes which lie near or at the point of attachment behave for all practical purposes as linked factors, for they may even be one unit apart. But then the difficulty arises, why these two non-homologous chromosomes remain more often in male than in female gametes. Cytologically too, no data on hand support this view.

HUXLEY furthermore was able to show by suitable selection ex-

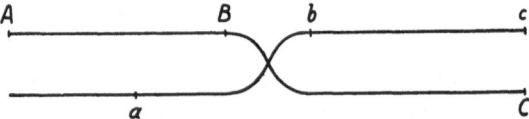

Fig. 2. From this figure it will be seen that A B C, and a b c
are non-crossovers while A B b c, and a C are crossovers.

periments that the great variability of the c.o.v. (9—25 %) among different males has a genetic basis. The possibility even exists that the loci do not correspond, which fact may lead to extraordinary results viz. reduplication and deficiency. (See fig. 2).

Concluding our discussion of HUXLEY's experiments on *Gammarus chevreuxi* it is apparent that crossing-over is more frequent in the female than in the male, while for the latter a greater variability of the c.o.v. was observed.

## § 7.  *Silkworm*

Another class of insects, which may be represented in this connection is that of the *Lepidoptera*, and especial attention will be given to the silkworm, *Bombyx mori*, studied by TANAKA (121). These experiments were intentionally carried out with a view of gaining insight in the "occurrence of systems of gametic reduplication in male and female hybrids". The results obtained up to 1913 all point to the conclusion that the "reduplication" between certain factors follows distinctly different systems in both sexes, a low partial taking place in the male while complete reduplication exists in the female". The experiments of TANAKA involved the following factors: 1. colours, Y = yellow cocoon; y = white cocoon; 2. markings, (a) S = striped or striped back; s = not striped; (b) M = moricaud; m = not moricaud; (c) N = normal; n = not normal.

We will briefly discuss the results obtained from the cross nyny ♀ × NynY ♂ and the cross NynY ♀ × nyny ♂. The former cross will show crossing-over for male gametes and the latter that for female gametes. For the cross nyny ♀ × NynY ♂ the results of the matings H 43.1 and H 43.2, were somewhat different as far as repulsion is concerned, although the female and the male parents were of the same fraternity, but n was not so large for the respective cases. Still we must admit that the ratio obtained agreed very well with the theoretical expectation as is apparent from a summary of this cross given by TANAKA. Repulsion, was very weak, for either system, $1:3:3:1$ and $1:2:2:1$. All the combinations were obtained, e.g. normal yellow, normal white, plain yellow and plain white. As far as the cross NynY ♀ × nyny ♂ (test for c.o. in female) is concerned only normal whites and plain yellows were obtained which means that in the female silk worm the repulsion between N and Y is complete i.e. no crossing-over occurs in the female between N and y or Y and n, (n was 5046).

For the backcross coupling experiment, concerning moricaud and yellow with the male parent heterozygous a gametic ratio of $3 : 1 : 1 : 3$ was obtained, and although the data from the reciprocal cross MYmy $\male$ × mymy $\female$ was not yet on hand, still the $F_2$ data obtained from $F_1$ moricaud yellows derived from the cross between moricaud yellow and normal or plain white, may be reasonably explained on the assumption (a) that complete coupling between M and Y is caused in ovogenesis of $F_1$ moricaud yellows (MYmy) and (b) that the male gametic ratios are $3 : 1 : 1 : 3$, i.e. a low coupling between M and Y. Thus from the gametic ratios, 3 MY 1 My 1 mY 3 my × 1 MY 1 my, we obtain 11 MY 1 My 1 mY 3 my phenotypes, which fits the actual facts found very well. In *Bombyx mori* therefore the results show no crossing-over in the female, while for the factors discussed crossing-over takes place very frequently in the male.

### § 8.   *Vertebrates*

We may pass on to discuss the crossing-over differences between male and female birds. COLE (23) studied 2 sex-linked characters in the pigeon. I is an intensifier of both black and red pigmentation in the feathers, and A is a modifier of black pigment. For the male pigeon crossing-over between these factors was found to be 40 %, while there was no crossing-over in the female. But in connection with the other experiments on the pigeon and the fowl it seems to me as if COLE's experiments only represent a special case, for CHRISTIE and WRIEDT (26) found crossing-over to occur in both sexes with perhaps the same frequency, e.g. crossing-over between magpie and shield-pattern was found to be 27.6 % in the female and 29.7 in the male pigeon. DUNN found the same for the fowl. Cytologically also, the impression gained from WILSON (The cell in heredity and development, 1925, p. 786) is that the results are not absolutely sure as yet.

The experimental results on rodents remain to be discussed. In this connection it will be noted that more ground is gained for generalisation, at least for rats and mice.

One of the most recent investigations as regards linkage and crossing-over in the house mouse is that of SNELL (103). He studied the linkage relations between hairless and piebald, and observed a difference between the two sexes. For the male a c.o.v. of 2.6 % and for the female a c.o.v. of 9.8 % was obtained.

CASTLE and WACHTER (22) carried out extensive experiments in order to test the linkage relations between the three complementary factors, albinism (c), red-eyed yellow (r), and pink-eyed yellow (p) in the common rat *Mus norvegicus*. The c.o.v. calculated on male and female gametes for albinism and pink-eyed yellow was found to be 19.67 % ± 0.26 and between yellow and pink-eyed yellow 18.33 % ± 0.69. Albinism and red-eyed yellow were strongly linked. The linear order was c-r-p. But crossing-over was found to be different in the two sexes the value in the female being the higher for each of these regions.

For c-r the c.o.v. for females was   0.53 ± 0.78; for males   0.18 ± 0.5
 ,, c-p ,,    ,,   ,,     ,,     ,,   21.93 ± 0.44 ,,    ,,    18.39 ± 0.32
 ,, r-p ,,    ,,   ,,     ,,     ,,   20.46 ± 0.92 ,,    ,,    15.55 ± 1.04

Their data obtained from experiments with mice were essentially of the same character as the foregoing. The c.o.v. for albino and pink-eyed coloured mice was 16.44 % ± 0.82 for females and 13.77 ± 0.5 for males, a difference of 3 units in favour of the females.

DETLEFSEN and CLEMENTE (31) observed the same phenomenon in their experiments on mice, besides, a more stabile value was obtained for the female gametes. It seems therefore that for mice it is a general rule that crossing-over takes place more frequently in females than in males. DETLEFSEN (30) however is of opinion that this difference may not in the first place be due to sex. He points out that prenatal elimination in certain cases may to all appearances cause a decrease in crossing-over, for the pink-eyed phenotypes are more liable to suffer. Consequently in the case of a coupling experiment the pink-eyed individuals suffer the most thereby reducing the netto c.o.v. But if sex does play its part then we should expect to observe a reduced class for the females of this coupling experiment as well.

It is not clear to me how viability affects the conclusion in this case, for we simply compare the pink-eyed phenotypes of both sexes. But DETLEFSEN goes still further and tries to point out that the sex differences in the c.o.v. so far obtained are not statistically significant, and he is inclined to ascribe these differences as due to random sampling. But one may ask the question why do all these differences obtained from the experiments carried out so far deviate in the same direction? DETLEFSEN's statistical treatment was therefore incomplete, and at the same time it may serve as a warning not to abuse the Probable Error. The Probable Error should not mar our results, for a difference may be very small and nevertheless be real. It is the good work of CASTLE (20) to point out how DETLEFSEN does not duly consider the force of 'cumulative evidence'. CASTLE reminds him of 9 different back-cross experiments carried out on linkage in rats and mice, none of these experiments counting less than 1000, one even counting 30000 individuals. Each of these experiments show a higher c.o.v. for the female heterozygous parent. A tenth experiment comprising more than 12000 young showed a c.o.v. between albinism and red-eyed yellow in rats three times larger in the female than in the male. No further comment is necessary, for anyone who reads DETLEFSEN's article: "The linkage of dark eye and color in mice" (30) will notice this mistake of individualising the P.E. There remains no doubt that crossing-over in rats and mice takes place more frequently in the female than in the male.

As regards difference in frequency of crossing-over between the two sexes in rabbits CASTLE (21) found that male heterozygous parents (English spotting and Angora coat, long hair) back-crossed produced 950 young with a c.o.v. of 14.4 $\pm$ 1.09; the female parents produced 609 young with a c.o.v. of 12.3 $\pm$ 1.37 so that apparently there is a smaller percentage of crossing-over in the female. But the difference is 2.1 % $\pm$ 1.75, which does not greatly exceed the P.E. and may therefore be a statistically doubtful difference. Yet it is remarkable that the same evidence is obtained for the Angora and the Dutch spotting back-cross.

CASTLE remarks: "The matter merits further investigation but the evidence already in hand would not lead us to expect to find any uniform sex difference in linkage throughout the class of mammals". However, as regards linkage values between chinchilla or one of its several allelomorphs and brown, we learn from CASTLE that $F_1$ females which produced 745 young gave 271 crossovers giving a c.o.v. of 35.9 $\pm$ 1.2, while $F_1$ males produced 255 young of which 75 or 29.8 % $\pm$ 2.1 were crossovers. In this case therefore the females have given a higher c.o.v. agreeing with the results obtained for rats and mice. The difference between the sexes is 6.1 with a P.E. of 2.4 i.e.

the difference amounts to $2.5 \times$ P.E., which is still a somewhat doubtful difference.

I think it will be more surveyable to make a summary of the cases

TABLE 5. Summary of cases of crossing-over in male and female compared.

| object studied | number of chromosomes etc. | name of cytologist | remarks pertaining to sex-difference in crossing-over |
|---|---|---|---|
| *Zea mays* | 10 hapl., no sex diff. | KUWADA | c.o.v. in the one case lower, in the other higher for the same sex, but diff. nearly always observed. |
| *Pisum sativum* | 7 hapl., no sex diff. | CANNON SAKAMURA | equal frequency for both sexes. |
| *Primula sinensis* | 9 hapl., no sex diff. | GREGORY | c.o.v. higher in ♀ for the one region, but higher in ♂ for the other region of same chromosome. |
| *Pharbitis Nil* | 14 hapl. | IMAI | c.o. perhaps with equal frequency in macro- and microsporogenesis. |
| *Drosophila melanogaster* and *D. simulans* | 4 hapl., ♂ XY type | BĚLAŘ and others | except for a few doubtful cases no crossing-over in the ♂. |
| *Tettigidea parvipennis, Paratettix* | for all species 6 hapl. in ♂ 7 hapl. in ♀ | ROBERTSON | c.o.v. in ♀ higher than in ♂. |
| *Bombyx mori* | 28 hapl. ♀ heterogametic | YATSU | no. c.o. in the ♀. |
| *Gammarus chevreuxi* | ? ♂ heterogametic | ? | c.o.v. higher in ♀ than in ♂. |
| *Columba livia domestica* | 8 hapl. ♀ heterogametic | HARPER | contrary results, COLE and KELLY: c.o. only in ♂, CHRISTIE and WRIEDT no definite difference in frequency between ♀ and ♂. |
| *Gallus domesticus* | 9 hapl. ♀ heterogametic | GUYER | uncertain, c.o. perhaps with equal frequency in ♀ and ♂ according to DUNN. |
| *Mus norvegicus* | 18 & 19 hapl., ♂ heterogametic | ALLEN | decidedly higher c.o.v. for ♀ than for ♂, pertaining to both rats and mice. |
| *Lepus cuniculus* | 11 hapl. for ♂ and ♀, ♂ heterogametic | BACHHUBER | somewhat uncertain; in some cases c.o.v. higher for ♂, in other c.o.v. higher for ♀. |

discussed above and thereafter to try to draw a general conclusion as far as the facts allow for such. In doing so the number of chromosomes will be accounted for, which intentionally was not mentioned in the above text.

It is very difficult to find a general rule as is apparent from this summary. For *Drosophila, Tettigidea, Acrididae, Locustidae,* and rodents, the male is heterogametic and for all these cases the c.o.v. is decidedly lower in the male, but for an uncertain case in connection with the rabbit. For *Bombyx mori,* crossing-over fails in the female altogether, she being heterogametic (WZ type). Unfortunately however, neither genetically nor cytologically may this be said of birds (pigeon and domestic fowl). Most probably the female is heterogametic (WZ type); genetically COLE found no crossing-over in the ♀ pigeon but CHRISTIE and WRIEDT are of opinion that crossing-over occurs with the same frequency in both male and female pigeon. Granting HUXLEY the benefit of the doubtful cases, it seems to me that his statement (1928) is logically the most cautious: "It appears to be a general rule that where-ever crossing-over is absent or markedly reduced in one sex, that sex is the heterogametic sex". But perhaps we may add to his statement that crossing-over does not definitely favour any sex where the male and female gametes do not differ cytologically as to the sex chromosomes, e.g. maize, *Pisum,* and *Primula.*

## CHAPTER VI

### CROSSING-OVER BETWEEN THE X- AND THE Y-CHROMOSOME

It is perhaps not too much to say that the absence of crossing-over between the X- and Y-chromosome of the male *Drosophila* really offers no problem for the simple reason that the morphological difference between the two chromosomes offers sufficient basis of explanation. But the cases that call for attention are especially those where either genetically or cytologically absence of crossing-over is not met with or not expected because of morphological and other reasons. Such cases as for instance crossing-over between the W- and the Z-chromosome of *Platypoecilus,* and crossing-over between the X and Y of *Aplocheilus latipes* and of *Lebistes reticulatus,* are cases we have in mind, especially because of the difference in frequency between the two directions of crossing-over.

FRASER and GORDON studied "The Genetics of *Platypoecilus*" and especially with reference to "The linkage of two sex-linked characters" (44). For this *Cyprinodont* it was found among others that the sex-linked factors "red" and "spots" are linked and liable to crossing-over. The dominant factors red and spots in the red race are as a rule ·located in the Z-chromosome while the recessive allelomorphs are as a rule borne by the W-chromosome. It is as yet not certain whether crossing-over takes place between the ZZ of the male. And the data referring to difference in crossing-over from W to Z, and vice versa are not thus statistically significant as to allow for a definite conclusion. Nevertheless, the data on hand makes it sure that crossing-over does take place between the W- and Z-chromosomes of *Platypoecilus*.

More definite results were obtained by AIDA, working with *Aplocheilus latipes*. AIDA (1;2) found that the sex-chromosome of *Aplocheilus* are of the *Drosophila* type, X Y. Furthermore he found that the factor R (red) is sex-limited, being borne by the Y-chromosome. Occasionally it may happen that R crosses over from Y to X and vice versa. It was interesting to know whether the frequency of crossing-over was the same in both directions. Consequently extensive experiments were commenced by AIDA (2) in order to determine these respective frequencies. In order to determine the frequency of transition of R from Y to X a heterozygous red male XrYR was mated to a homozygous white female XrXr and it was found that the frequency was 1 : 300 approximately. To determine frequency of transition of R from X to Y a heterozygous red male XRYr was mated to a homozygous white female XrXr. The frequency was found to be 1 : 1200. From this total the non-disjunctional males and females were excluded.

The gene R therefore transits from Y to X with a frequency ratio 1 : 300 and it transits from X to Y with a frequency ratio of 1 : 1200 which means that it crosses over from Y to X with a frequency four times that of transition from X to Y. AIDA states the following: "If the crossing-over between the X- and the Y-chromosome takes place simply by chance and no other cause interferes with it, the frequencies of transfer of the R gene in both directions should be equal, and no preponderance on either side should take place. At present I am unable to elucidate the cause of this difference in the direction of crossing-over. Whether it is effected by a dissimilarity in structure of the X-

and Y-chromosomes or by the action of some gene or genes situated in one of them is not known. However the tendency of the Y-chromosome more frequently to lose the dominant gene and less often to regain it is well established in our own fish and also in *Lebistes*; and if this difference is supposed to occur also in other animals we have here perhaps a plausible explanation of the origin of the socalled empty Y-chromosome of *Drosophila* where crossing-over must have taken place so repeatedly between the X- and Y-chromosomes that the Y-chromosome has gradually lost its dominant genes eventually to become quite empty".

Although STERN has succeeded in localizing the factors male fertility, $K_1$ and $K_2$, and "langborstigkeit", bb in the Y-chromosome still it remains relatively empty, for all that.

But it would have been noted that AIDA meant to have found non-disjunction in the male *Aplocheilus*, in contradistinction to non-disjunction in the female of *Drosophila*. At reduction division homo-synapsis would then take place more frequently in non-disjunctional *Aplocheilus* males. The lagging Y may altogether fail to enter the nucleus, but in very exceptional cases X and Y may enter the nucleus together thus giving secondary non-disjunctional males. It must be pointed out that if such is the case then non-disjunction happens more often in *Aplocheilus* than in *Drosophila*. WINGE (136) points out that in order to explain the great majority of females derived from exceptional males in *Aplocheilus* it must be assumed that the two X-chromosomes fail to seperate after conjugation prior to reduction division. The Y-chromosome then remains "as a superfluous solitary chromosome lagging in the middle of the spindle possibly to be excluded from the formation of a gametic nuclei". WINGE added: "AIDA is evidently disturbed by the presence of the Y-chromosome". WINGE would like to see the results tested out as was done with the results on *Lebistes*. With *Lebistes* it is possible to bring forth XX males, which produce only female offspring. There is every reason to believe the *Aplocheilus* males in question to be of the same nature. Cytological investigation might decide the question. The explanation of $X_oX_o$ males may according to WINGE be explained as follows. Both X and Y bear factors for colouring. The difference between X and Y is that Y bears a dominant male determining gene, while X bears a recessive female determining factor or none at all. This gene is absolutely Y-linked. This is the case with respect to the gene maculatus in *Lebistes*, which may be looked upon as identical with the male determining gene. WINGE found that females which show indications to change into the male direction, are homo- or heterozygous for coccineus (Co) and vitellinius (Vi) in the X-chromosome. Also such females bearing the genes luteus (Lu) and tigrinus (Ti) in the X-chromosome and such as showed signs of developing in the male direction might sometimes show the elongatus factor (El = elongated caudal fin).

An interesting case was the following:

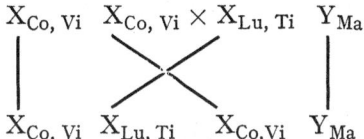

$$X_{Co, Vi} \quad X_{Co, Vi} \times X_{Lu, Ti} \quad Y_{Ma}$$

$$X_{Co, Vi} \quad X_{Lu, Ti} \quad X_{Co, Vi} \quad Y_{Ma}$$

In this case it was to be expected that all three colours Co, Vi, and Ma would show in the males. Actually 55 out of the 58 cases were according to expectation, but 3 males did not show the character for maculatus and were coccineus, vitellinius, luteus and tigrinus, which was the expected formula for the daughters. The explanation offered was that the three males were XX males of the constitution $X_{Co, Vi}$ $X_{Lu, Ti}$.

They were crossed to $X_o X_o$ females and the result was that the offspring amounting in all to 314 were all females as follows:

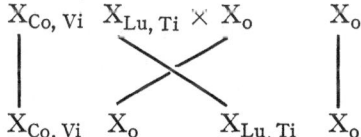

$$X_{Co, Vi} \quad X_{Lu, Ti} \times X_o \quad X_o$$

$$X_{Co, Vi} \quad X_o \quad X_{Lu, Ti} \quad X_o$$

On the basis of WINGE's suggestion AIDA's results may be explained thus: heterozygous red *Aplocheilus* males, XRYr × white females XrXr from which were expected the genotypes XRXr and XrYr; but observed were 5 white females XrXr and 9 red males, XRXr. The latter were specimens which, notwithstanding their chromosomal constitution XX, still developed in male direction "in consequence of a casual accumulation in the X-chromosome and in the autosomes of genes with a faintly female and stronger male effect respectively. Normal sex determination depending on XX or XY constitution is here disturbed, other genes having taken the lead in the sex determination".

WINGE therefore is far on the way of accepting a quantitative theory of sex determination in *Teleosts*, the amount of crossing-over and accumulation of genes from the usual chromosome to the other being the agencies at work. This explanation suffices for all possible intermediate stages.

As to the frequency of crossing-over from X to Y and vice versa, WINGE's results on *Lebistes* also seem to point in the direction that for this fish crossing-over from X is different from that from Y. The case is an interesting one, the point at issue being the adaptation of the chromosome which receives the gene to that chromosome from which it exits. The problem will be further discussed in Chapter I, Part II.

## CHAPTER VII

### NON-DISJUNCTION, POLYPLOIDY AND RELATED PHENOMENA AND THEIR EFFECT ON CROSSING-OVER

The phenomena of non-disjunction and attachment of X-chromosomes were of great interest to MORGAN's school, for it offered invaluable evidence, especially as far as sex-linked genes are concerned, for the chromosome theory of heredity. The behaviour of the genes could not be explained in any other way. Of lately these phenomena came still more in the focus of attention for two reasons. It appears possible to a fairly high degree to produce these phenomena artificially, e.g. by means of X-rays (MAVOR: 70, 71). But also cytogenetics profited a great deal by this abnormal behaviour of the sex-chromosomes. Cytologically, it has a great advantage over translocation in so far as attached chromosomes and an additional chromosome are relatively easier to be seen.

Non-disjunction, attached chromosomes and polyploidy all have a bearing on the problem of crossing-over. Several investigations are at our disposal, and they have led to important discoveries. They are of a fairly recent date. As far as non-disjunction is concerned we may refer to ANDERSON's (5) "Studies on a case of high non-disjunction in *Drosophila melanogaster*". By means of X-ray treatment ANDERSON obtained among a group of primary non-disjunctional females, one which gave an exceptionally high ratio of secondary exceptions, amounting to more or less 23 % as compared with 3 to 4 % for the normal stock. The X-chromosome of the females of the exceptional stock was of the constitution $\dfrac{+\ +\ cv\ +\ v\ +\ f}{sc\ ec\ +\ ct\ +\ g\ +}$ (sc = scute 0, ec = echinus 5.5, cv = crossveinless 13.7, ct = cut 20.0, v = vermilion 33.0, g = garnet 44.4, f = forked 56.5, the numbers signify the respective loci in the chromosome map).

The high percentage of non-disjunction was due to a gene in the cv-v-f chromosome, located near to vermilion which pleiotropically as it were, also affected the c.o.v. Crossing-over for instance was normal for the region from sc to cv, but to the right of cv crossing-over was strongly reduced, calculation being based on regular males. Also the total amount of crossing-over between sc and f was only 20.5 % in regular XX females heterozygous for the high non-disjunction gene as contrasted with 62.2 % in the controls.

Experiments were carried out to test the chromosome which was free from the factor for high non-disjunction, and c.o.v.'s were obtained which agreed with the standard map. It was found furthermore that crossing-over in high non-disjunctional regular XX females was undoubtedly lower than in regular XXY females of corresponding constitution. Crossing-over between sc and f for instance was 20.5 % in XX females, but 29.0 % in XXY females, a ratio of 0.71 : 1. Crossing-over from secondary exceptions was much lower than that obtained from regular females, both cases referring to calculations based on progeny of the XXY high non-disjunctional females.

The reduction in the c.o.v. for secondary exceptions is especially remarkable for the right end of the chromosome.

If the calculation is based on the exceptional and the regular XXY's, the total of all the c.o.v.'s from sc to f amounts to 20.9 according to ANDERSON, which amount does not differ much from that obtained in regular offspring of XX females, viz. 20.5 %. The total percentage for XX females was not altered much when the primary exceptions, XXY females were included. This great resemblance between the c.o.v. totals for XXY and XX females suggests that crossing-over frequency is the same for XXY and XX females, so that the presence or absence of Y has no influence on the c.o.v.

With reference to the above, I think it fit to remind the reader of MAVOR's (73) experiments with respect to the effect on crossing-over and non-disjunction of X-raying the anterior and posterior halves of

*Drosophila* pupae. His results were, that only X-raying the posterior part of the pupae was of real consequence as to its effect on crossing-over and non-disjunction. Consequently only direct X-raying of the germcells raised the percentage of non-disjunction with a consequent change in the c.o.v.

Closely related to the phenomenon of non-disjunction is that of attached X-chromosomes, and it would be interesting to know how the c.o.v. is affected in this case. In a preliminary X-ray test carried out in 1924 on $F_1$ females derived from the cross $\frac{\text{sc wa cv t f}}{\text{sc wa cv t f}} \times$ br ec ct g, ANDERSON (4) obtained among the non-disjunctional females, one which gave only exceptional offspring. Phenotypically she was a broad echinis and her two X-chromosomes which were attached at the forked end were of the constitution $\frac{\text{br ec cv t f}}{\text{br ec cv t f}}$, (sc = scute, br = broad, wa = apricot, t = tan). Mated to a Bar male, only broad echinus females and Bar males were obtained. Among the broad echinus females however there were some which showed additional recessive characters viz., cv, ct, t, g and f which characters were present in heterozygous condition in the broad echinus female parent. Here therefore, an opportunity was offered to study crossing-over between the two attached X-chromosomes of the heterozygous mother. From this mother 138 daughters were identified, and the c.o.v.'s calculated for the different regions. It was remarkable to find that the c.o.v.'s were similar to, or well within the range of the standard values of normal non-attached X-chromosomes.

Considering the small numbers involved the values obtained for the attached chromosome agree fairly well with standard expectations. Thus the data show that crossing-over takes place normally even though the chromosomes are attached.

Occasionally the two attached chromosomes become detached. The c.o.v.'s from these regular females are probably similar to those of the exceptional daughters.

A few months previous to ANDERSON's experiment mentioned above L. V. MORGAN (77) studied a case of polyploidy in *Drosophila* with two attached chromosomes. From a mosaic of *Drosophila melanogaster* with a yellow female region where the two X-chromosomes were attached to each other by their ends, a socalled double yellow stock was maintained. These two X-chromosomes carried the gene for yellow body colour, and since they remained attached, the stock consisted of yellow females derived from eggs with attached X-chromosomes fertilized by a Y-chromosome sperm and of males which were like their father derived from eggs with a Y-chromosome fertilized by an X sperm, which two classes namely XXY females and XY males were represented in equal proportions. Occasionally XX X females occurred, being derived from an XX egg and an X sperm. These XX X females were easily identified morphologically.

Thus the opportunity presented itself to study crossing-over in the case of attached chromosomes with respect to XXY females and XX X females. It was also studied in 3n females. It was further possible to study the effect on crossing-over in the autosomes in cases of triploidy

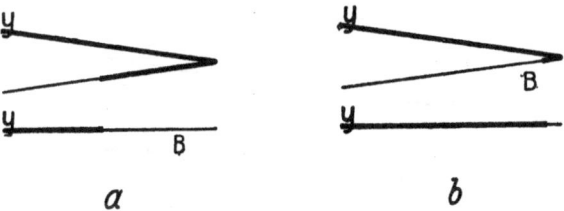

Fig. 3. Two possible crossovers between two attached chromosomes, both carrying yellow, and a free chromosome carrying Bar. After L. V. Morgan (77).

as well as in the cases of attached X-chromosomes. The attached X-chromosomes were of identical constitution. As far as the possible effect in the autosomes is concerned, a black, purple, vestigial, yellow heterozygous female was mated to a recessive male. The data obtained showed that crossing-over took place normally for these regions in the second chromosome.

As far as the third chromosome is concerned the general impression is the same as that got from the second chromosome, but the data are somewhat less convincing.

It was not possible as a matter of course to study crossing-over in the attached chromosomes, for they were of identical constitution. Nevertheless this was done in other experiments. Into this double yellow stock was brought a complete set of chromosomes (culture 182 STURTEVANT) of which the X-chromosome did not carry the yellow gene, but carried the genes + and Bar. It thus became possible to study crossing-over between the attached X-chromosomes and the free X-chromosome. The $F_1$ males from STURTEVANT's culture No. 182 consisted of two classes namely, Bar (non-crossovers) and yellow Bar (crossovers), but the usual complementary class (wild) was wanting. The explanation is that the free chromosome crossed over with one of the attached chromosomes. As is apparent from Fig. 3, the Bar bearing free chromosome crossed over with one of the attached chromosomes resulting in a free chromosome of yellow Bar or one of yellow wild. Only non-crossovers (Bar) and crossovers (yellow Bar) were observed. Further experiments affirmed this result.

However, with this case of crossing-over between the free X-chromosome and one of the attached chromosomes the latter of

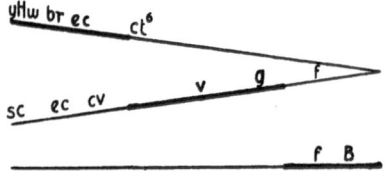

Fig. 4.    Diagram of crossing-over among
three chromosomes. After L. V. MORGAN
(77).

course became different from its mate to which it was attached at the right end. In this manner a 3n female was obtained in which one of the attached chromosomes carried the gene for yellow and the other differentials for scute, echinus, crossveinless, vermilion, garnet and forked, the combination being called "Z-ple". A wild phenotype, 2n non-disjunctional daughter from an egg with attached X-chromosomes of above mentioned constitution, gave in the $F_1$, wild type daughters, yellow daughters and daughters with different combinations of the mutant characters. By further breeding of the wild type females it became possible to observe crossing-over in the attached X-chromosomes. It also happened that a crossover chromosome was made up of parts derived from all three chromosomes. Thus a yellow, hairy wing, broad, vermilion, garnet, forked, Bar male was obtained which most probably also possessed the gene for echinus. (This latter character however is difficult to differentiate when Bar is present.) The explanation was as follows: the same result would be obtained when crossing-over occurred in the attached chromosomes between the loci for crossveinless and cut, and further if crossing-over with the free chromosome occurred at a point between the loci for garnet and forked, as is evident from Fig. 4.

These experiments therefore, prove that crossing-over may take place ,in triploid *Drosophila* females (a) between the attached X-chromosomes and the free chromosome and (b) between the two attached X-chromosomes themselves, and (c) mosaically, between the attached X-chromosomes and the free X-chromosome inter se.

Furthermore, it was shown that crossing-over occurred in the autosomes of the females with attached X-chromosomes, at least this phenomenon was observed in connection with crossing-over in the second and the third chromosomes of the females in question.

BRIDGES and ANDERSON (18) however, considered the possibility that the distribution of the c.o.v.'s over the X-chromosome might be different in the case of diploid and in the case of triploid *Drosophila* females. They made use of control tests and made sure that this possible difference was not due to unknown crossover modifiers brought in with the special X-chromosome to be used. They obtained data as to the crossover values from the same chromosomes after they had passed through the triploid stage into the ordinary diploid stage. From the data it was apparent that the number of recombinations in the first section as regards triploid females was more than twice as large as that of the diploid control females. Furthermore it appeared that in the triploid females there existed great regional differences as regards the deviation from the standard crossover values of the X-chromosome. This means that a unit of crossing-over in the extreme left end of the X-chromosome map, may correspond to a longer section than the unit of crossing-over in the region more to the right thereof. It is furthermore pointed out by BRIDGES and ANDERSON (18) that the frequency of mutation per map unit is relatively higher in the left end of the X-chromosome.

These, undoubtedly, are invaluable results but to my mind the results obtained from the 'aberation' studies (translocations etc.) offer a still more suitable means to 'correct' our map, for we may thus obtain better insight in the regional differences of the chromosomes, which on its turn means gaining more insight in the physiology and physics, not only of crossing-over as such but also of the general basis of genetics. This does not mean that we are inclined to give values to scientific results, for scientific results are unique for selfevident reasons and therefore incomparable as such. But it may happen that in a certain phase of a science a certain problem presses for solution and then of course one may say that for this special case these results, are of more or less importance. This conviction made me attach special importance to experiments where the process of crossing-over may possibly be vigorously interfered with, e.g. the effects of X-rays, radium, and possibly the effects of ultra-violet rays, and of violent centrifugal force.

## CHAPTER VIII

### CHANGES IN THE CROSSING-OVER VALUE DUE TO TRANSLOCATION AND RELATED PHENOMENA

The class of chromosomal disturbances discussed in the former chapter comprises those concerning whole chromosomes or groups of whole chromosomes. Polyploidy and non-disjunction are the phenomena most commonly observed in this connection. Here the constellation of the chromosome group is interfered with. The class to be discussed in this chapter concerns actual aberations of sections of chromosomes; it comprises the phenomena of translocation, deficiency, inversion and duplication of a fragment or section of a chromosome.

The phenomenon of translocation was first observed in *Drosophila* by BRIDGES in 1917 and in 1919 referred to by MORGAN. In 1922/23 we find a preliminary account of it by BRIDGES himself in "The Anatomical Record" 24, 1922/23; p. 426/7: "The Translocation of a Section of Chromosome II upon Chromosome III in *Drosophila*". It concerns a mutant eye-colour, pale, which was considered to be a non-sex linked specific modifier of eosine without effect by itself. Crossing-over experiments seemed to show that this modifier, P II, was in fact located in the whole section measuring some 8 units, from arc to the right end of the second chromosome, in which section crossing-over was entirely suppressed. A similar modifier, P III, located to the left of rough suppressed nearly all crossing-over from spineless to the right end of chromosome III, i.e. for nearly 50 units or nearly half the whole length of that chromosome. In the presence of P III all the recessive mutant characters viz. plexus, brown, speck, morulla, balloon, blistered, purpleoid and lethal IIa which are located to the right of arc in chromosome II, failed to show even when present in homozygous condition. It was only natural to interpret P III as a broken off end of chromosome II, and P II as chromosome II from which this end has been broken off.

At the instigation of MULLER, DEBUZ HAMLETT (57) tried to throw more light on the linkage relations of chromosome III, to which the broken off part of chromosome II was attached. But whereas BRIDGES only studied the case in heterozygous condition, HAMLETT also did it in homozygous condition. Crossing-over throughout the whole left end of this chromosome was exami-  ned. It was found that crossing-over at the place of attachment of the fragment was reduced in homo- zygous condition as well, so that interpolation of this fragment seemed improbable, for one would ere ex- pect the c.o.v. to rise in homozygous condition of the fragment. It was concluded that chromosome III was branched at the place of attachment causing a reduction of crossing-over. HAMLETT could show further that crossing-over in this case was not redu- ced over the whole chromosome. Starting from the left, crossing-over remained normal till the locus of pink was reached, i.e. a third of the length of the chromosome. From here the map distance was re- markably shortened till the locus of the translocated fragment was reached, where crossing-over was about 1/3 normal. Between rough and claret crossing-over rose again gaining half its normal value.

Fg. 5. Transloca- tion of a piece of the X-chromosome to one of the third chromosomes. Af- ter T. S. PAINTER and H. J. MULLER (86).

Translocation may be brought about by physical agents e.g. X-rays. MULLER (1926) was the first to produce translocation by means of X-rays, that is to say translocation which could be tested out genetically and cytologically. The X-chromosome was broken and the fragment became attached to the third chromosome.

Further may be mentioned PAINTER and MULLER's (86) "transloca- tion III to II 26" (Fig. 6) which was also studied genetically and cyto- logically. Cytologically they were able to show that the translocated fragment was not as long as the length arrived at genetically. Gene- tically, for instance, the break occurred between the locus for stripe at 62 and that of ebony at 70.7, so that the loss of the third chromosome amounted to at least 35 units, that is to say about 1/3 the length of 106. 2 units of the genetic chromosome. Further genetic data showed that this right end or lower fragment of chromosome III was attached to the left end of chromosome II, and that the point of attachment of the fragment was not at its ordinary terminal point but at its broken off end. When these chromosomes were studied cytologically it became clear that the fragment in question was very small and difficult to see. Measurement of the differences in length between the homologous second chromosomes and between the homologous third chromosomes showed that the fragment which was exchanged, measured far less than 1/3 of chromosome III.

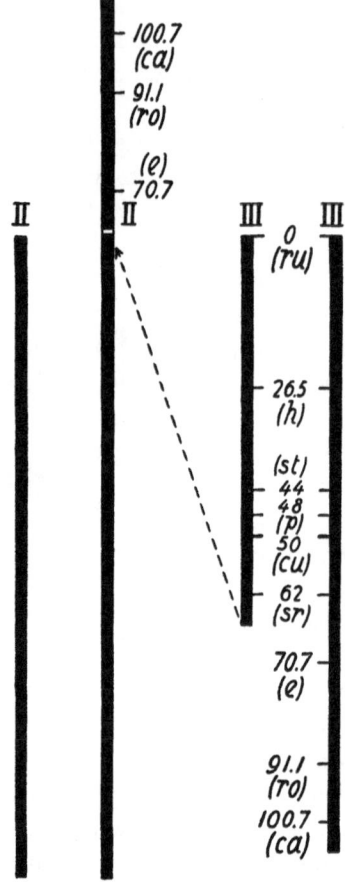

Fig. 6.  A "III to II" translocation. In this case it is the "right hand" end of the third chromosome which has broken and become attached to the "left hand" end of the second chromosome. After T. S. PAINTER and H. J. MULLER (86).

Other important cases of the same order were the following cases of deletion. When *Drosophila* males were X-rayed and then crossed with females from a stock with an attached X-chromosome, it might happen that in 1 % of the cases the X-chromosome might undergo deletion of the mid-region, a small fragment of the left end becoming attached to a small fragment of the right end which carried the spindle-fibre. Such a chromosome in conjugation with the attached X-chromosomes bearing yellow, resulted in a non-yellow (gray) female. Several deletions of the X-chromosome of this general type were analysed and it was found that the cytological size of the surviving fragment differed from its genetical size. As regards such deletions 13 loci were tested genetically in all the cases. Nearest to the left end (top end) were the loci for yellow, scute, broad, prune and white; nearest to the right (lower) end were the loci for carnation and bobbed. In the case of the socalled "Deleted X I" were present the loci for yellow, scute, broad and prune, but deeper inwards white was wanting in this left end of the chromosome. In the right end the locus for bobbed was present while carnation at 65.5 was altogether absent. None of the remaining loci lying in between was represented.

The same phenomenon was met with Deleted X2 and Deleted 14. The former (Deleted X2) chromosome was deleted from prune and white as far as bobbed. As regards Deleted 14 the loci for yellow and scute were present but deeper to the right none was represented till bobbed was reached at the extreme right end. Cytological examination showed that this deleted X-chromosome measured about 1/4 of the usual length, while genetically the map of the deleted X-chromosome did not show more than 1/12 its usual length. (The deleted types of the flies were phenotypically different).

These are very important facts especially when viewed from the point of correction of the c.o.v. as a function of chromosome distance. PAINTER and MULLER were convinced, as a result of their many investigations in this direction, that the combined study of genetical and cytological translocations and deletions would offer a fruitful method by means of which a true chromosome map may be constructed in agreement with the chromosome which we see through the microscope. Such a map would represent not only the true order of the genes, but also their actual spatial relations.

While several cases of disharmony between the genetical and the cytological map have been found and reported, one fact stood firm namely the order of the genes. This is a fact of great importance. Furthermore these translocation studies actually proved that whole series of genes were exchanged in linear order, thus also justifying the theory of crossing-over. But the problem begins when the genetical measurement is compared with the cytological. The c.o.v. functions for all practical purposes as the unit of measurement. But it may happen that the translocated fragment of a chromosome (PAINTER and MULLER (86), measuring more than 30 units in a sparse region may be very difficult to see under the microscope, while conversely a fairly long fragment, cytologically, may measure but a few units even in a dense region of the chromosome, which means that the scale of the chromosome map differs from region to region. This difficulty may perhaps be overcome by making use of the combined genetic and cytological results as treated above. It is possible that such a corrected map may be quite different from the present map and it may be more in agreement with the chromosome as seen through the microscope. DOBZHANSKY (33, 35) made a very interesting trial to construct such a corrected map for the 3rd chromosome in 1930 and

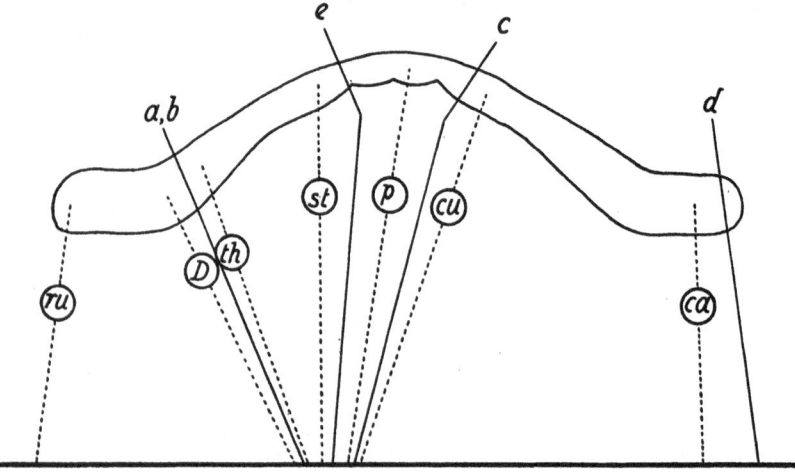

Fig. 7.   Comparison of a genetical (below) and a cytological (above) map of the third chromosome of *Drosophila melanogaster*. a, b, c, d and e-the observed breaking-points ru-roughoid; D-Dichaete; th-thread; st-scarlet; p-peach; cu-curled; ca-claret. After TH. DOBZHANSKY (33).

for the 2nd in 1931. That of the 3rd chromosome is here represented in Fig. 7.

How is this actually to be done? In the first place we must base our work on the changes in the c.o.v.'s observed, when translocation is involved.

For, although, by studying the *Drosophila* literature in connection with translocation, one is impressed by the many differences in the results obtained, yet one fact remains the same: whereever transloca-tion is concerned the frequency of crossing-over is reduced in the chromosome in question, or in the limb of the V-shaped chromosome in which the break occurs or in the limb of the chromosome to which the broken-off fragment is attached. The following cases may serve as illustration.

Reduction in crossing-over was observed by DOBZHANSKY (34) to occur in flies heterozygous for translocation. Such flies carry one chromosome broken into two parts and another sound homologous chromosome. In Fig. 8 the broken chromosome is the donor (A2). The fragment (A3) where the place of attachment of the spindle-fibre (X) is wanting became attached to the recipient chromosome B2. The remaining fragment of the donor (A2) which possesses the place of attachment of the spindle-fibre (X) remains free.

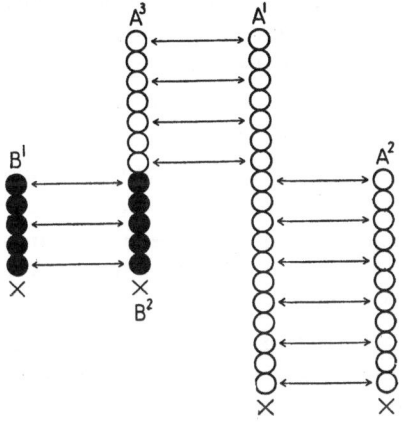

Crossing-over may now take place between each of the sections (A3, A2) of the broken chromosome, and the corresponding parts of the normal homologue (A1). Sometimes double crossing-over may take place simultaneously in both sections of the donor chromosome. From this it may be deduced that both fragments may conjugate simultaneously with the normal chromosome.

DOBZHANSKY studied 5 such cases of translocation with re-

Fig. 8. Scheme of attraction between chromosomes involved in a translocation. The donor chromosome and its homologue are represented in white, the recipient chromosome and its homologue are in black. The locus of the spindle-fiber is marked by X. The arrows indicate the direction of the attraction forces between the homologous loci in different chromosomes. After TH. DOBZHANSKY (34).

gard to the third and the fourth chromosome of *Drosophila*. In each of the cases a fragment of the third chromosome happened to break off and to attach itself to chromosome IV. In 3 cases where the third chromosome broke between the place of attachment of the spindle-fibre and the left limb of the chromosome, that is to say in the region roughoidpeach, crossing-over was observed to be considerably reduced. For the remaining regions right of the spindle-fibre crossing-over remains normal. Perhaps the latter regions even showed a small rise in crossing-over.

In two other cases the third chromosome was broken off to the right of the place of attachment of the spindle-fibre, with the result that crossing-over was reduced in the right limb.

DOBSHANSKY also studied 4 cases of translocation in which a section of the second chromosome was transferred to the fourth chromosome. In 2 of the cases the break occurred in the right limb of the second chromosome with the result that crossing-over was greatly reduced in this limb, while crossing-over remained normal in the left limb. The other 2 breaks occurred in the left limb, that is to say left of the place of attachment of the spindle-fibre, at the locus for pr, with the consequent decrease of crossing-over in this left limb while crossing-over involving loci to the right of the spindle-fibre, remained normal. It was a remarkable fact that the greatest reduction of crossing-over was near to the break of the chromosome, — in the pr-c interval for translocation a, and in the c-px interval for translocation d, both cases in connection with the 2nd chromosome.

Several other interesting cases in connection with chromosomes II and III were studied by DOBSHANSKY and STURTEVANT (119) in 1930. In 2 of the cases the 2nd and the 3rd chromosome were broken off at the place of attachment of the spindle-fibre, the left limbs of the 2nd and the 3rd chromosome then combining to form a new chromosome, the same applied to the right limbs. In the 3rd case the 3rd chromosome was broken off at the place of attachment of the spindle-fibre while the 2nd chromosome broke left to the place of attachment of the spindle-fibre. The right limb of the 3rd chromosome became attached to the longer fragment of the 2nd chromosome, while the shorter fragment of the latter became attached to the left limb of the 3rd chromosome. A significant decrease in crossing-over was observed in the left limb of the 2nd chromosome, some reduction of crossing-over throughout the 3rd chromosome, while the right limb of the 2nd chromosome gave normal c.o.v.'s.

In the 4th case a section of the left limb of the 2nd chromosome became attached to the left limb of the 3rd chromosome, but N.B. not to its end. The reduction in crossing-over was according to expectation: crossing-over was remarkably reduced in the left of the 2nd as well as of the 3rd chromosome while it remained normal for the right limbs for both the 2nd and the 3rd chromosome.

DOBZHANSKY also observed this phenomenon with regard to the X-chromosome and the 2nd chromosome, the former breaking at the locus for Bar and the latter at the locus for vestigial that is to say in the right limb. The broken-off parts of both chromosomes were exchanged. A reduction of crossing-over was observed in the right limb of the 2nd, and at least in a part of the 1st chromosome, while the left limb of the 2nd chromosome showed normal c.o.v.'s.

Ten further translocations with respect to the Y-chromosome and the 2nd chromosome were studied by DOBZHANSKY in 1930. In 5 of the cases relatively small parts of either the right or the left limb of the 2nd chromosome, including one of the ends, was broken and became attached to the Y-chromosome. Again crossing-over was reduced in the limb of the chromosome in which the break occurred, while the c.o.v.'s were fairly constant for the opposite limb.

BOLEN (12) in 1931 presented evidence of the first demonstrated case of breakage of the fourth chromosome produced by X-rays. The X-chromosome was broken between white and facet and the left end attached to the fourth chromosome. But breakage also occurred between eyeless and bent located on the spindle-fibre portion of chromosome IV and the left part containing eyeless became attached to the right portion of the broken X-chromosome. Unfortunately cytological data are still wanting. Summarising these cases we may say that when a break occurs between the spindle-fibre and the end of a chromosome, or when a fragment of another chromosome becomes attached to a limb of still another chromosome then a reduction of crossing-over takes place for that chromosome, the strongest reduction occurring at the place of attachment, or at the place of the break. I think that these results suggest still more namely that the homologous chromosomes should be fairly similar in length and that the loci should correspond as much as possible. And although DOBZHANSKY is of opinion that it is not the time yet to speculate as to the attraction of homologous chromosomes, still his tentative explanation seems to me a very plausible one, namely that for normal chromosomal conjugation with crossing-over taking place in the normal way, there should be an attraction between the homologous loci which should take place at a certain moment after gametogenesis, the attraction of homologous chromosomes being granted.

A new type of translocation in *Drosophila melanogaster* was studied by RHOADES (95) in 1931. He treated a wild type of Oregon *Drosophila* males with X-rays. After suitable crosses it was shown genetically that the X-rayed 2nd chromosome which had no recessive genes (wild), showed a linkage with the X-rayed Y-chromosome. Further analysis showed that a small fragment from the central region of the 2nd chromosome involving the loci for hooked (hk) purple and light (li) became attached to the Y-chromosome. The minimum of map distance of the translocated fragment measures 1.1 units while the maximum length measures 4.1 units. Cytologically however no difference in the length between the homologous of the 2nd chromosome, one partner of which was deficient for the translocated fragments, was observed. Furthermore very interesting results were obtained from the "duplication" females that is to say females with 2 normal 2nd chromosomes and an extra fragment from the 2nd chromosome, but attached to the Y-chromosome. A great reduction in crossing-over was observed in the left arm of the 2nd chromosome, while more or less normal c.o.v.'s were obtained for the right arm compared with the control. The greatest reduction was observed in the region which involved the translocation.

It might be that the Y-chromosome attached to the fragment of the 2nd chromosome, disturbed normal synapsis of the latter chromosome causing a reduction in the c.o.v.'s for the translocated females, for in the duplicated females crossing-over was reduced in the left arm of the 2nd chromosome.

ALEXANDER WEINSTEIN (127) in 1928 reported that among the offspring of X-ray treated *Drosophila* 6 cases were met with in which the genes of the 2nd chromosome behaved as if attached to the X-chromosome, that is they behaved as sex-linked genes. Unfortunately cytological data were not available. However it offered a case genetically similar to the former and it seems to me highly probable that it was a case of translocation too.

Closely related to these phenomena is the re-arrangement of genes. As an instance I may refer the reader to STURTEVANT's report in 1921 concerning a case of rearrangement of genes in *Drosophila*. STURTEVANT (114) pointed out as will be remembered that the crossover modifier CIIr and CIII reported by MULLER and STURTEVANT respectively may be nothing else but a re-arrangement of genes. "These 'genes' both cause in individuals heterozygous for them the disappearance of crossing-over in the immediate region where the genes themselves lie, and the considerable reduction of crossing-over in neighbouring regions. In individuals homozygous for either of these genes however, the percentage of crossing-over rises to or beyond that found in normal individuals. Experiments are now on the way in an attempt to determine if these genes are really simply inverted chromosome sections, but it will probably be a long task to settle the matter".

Another factor of the same category causing change in the c.o.v. is the phenomenon of deficiency. As an example we may take MOHR's (76) genetic and cytological analysis of a section deficiency involving 4 units of the X-chromosome in *Drosophila melanogaster*. It is perhaps the most extensive essay written on deficiency. The case was briefly as follows: a new mutation, Notch 8, in the X-chromosome affected a definite section near to the left end of the X-chromosome involving the loci for white, facet and Abnormal Abdomen, 1.5, 2.7 and 4.5 units respectively distant from yellow. Although the genes white and facet are recessive still they showed themselves in the heterozygous female, but crossing-over was completely suppressed in this region, for among 10294 females there did not occur a single case of crossing-over between white (eosine) and Notch 8.

In still another experiment Notch 8 heterozygous females were back-crossed to yellow white facet males. Although the standard c.o.v. between white and facet is 1.2 % none was observed in this experiment, nor was there any crossing-over between any of these two and Notch. Crossing-over however occurred between yellow and Notch in 1.7 % of the cases, which percentage agrees with the standard c.o.v. between yellow and white. The deficiency therefore started at white and reached 3 units further. Concerning the cytology, MOHR is of opinion from a careful study of all the cells, that the central ends of the X-chromosomes of Notch 8 females are slightly unlike each other while no such asymmetry is met with in the wild type.

The combined study of genetics and cytology in connection with translocation and related phenomena, and especially where it concerns the application of physical means (X-rays) to produce them, are achievements of most recent date. It opens a new era in the history of genetics, and with great expectation further developments in this direction are awaited.

PART II

CHAPTER I

*Introduction*

In connection with our experiments with *Lebistes* we had to devote the greater part of the time to the technique. From the start we closely followed the methods prescribed by WINGE (135) but in the course of our experience we brought about several alterations which seemed worthy of special mention. The method of preparing aquaria and the plants used were exactly the same as that prescribed by WINGE, so that this part of the technique will not be dealt with. Neither is it the intention to give any further description of *Lebistes* for this was fully done by WINGE, BLACHER (11, 12) and others. However, in connection with our observations special mention will be made of the development of the colour patterns of the male. In the paragraphs to follow the technique of air-supply, regulation of temperature, artificial illumination, feeding methods and diseases will be briefly dealt with.

I have great pleasure in expressing my great indebtedness towards Prof. WINGE, who kindly sent us the *Lebistes* material.

§ 1. *Air supply*

Especially during the winter months when twilight set in early and when sunshine was very scarce, $CO_2$ assimilation was too low to supply the aquaria with sufficient oxygen. We were consequently forced to supply air by artificial means. This was done by means of a K.D.A. airpump (see fig. 9) which was connected to the watertap (T) in the room. The pressure of the water amounted to more or less 3 atmospheres. The speed was regulated by turning the tap.

The K.D.A. airpump (from A. Glaschker, Taucheerstrasse 26, Leipzig, C 1) is constructed on the twin stroke principle and has a very great efficiency. Its water consumption is very low, being $\pm$ 9 litres per 11 hours, supplying air for 50 aquaria. There is hardly any wear

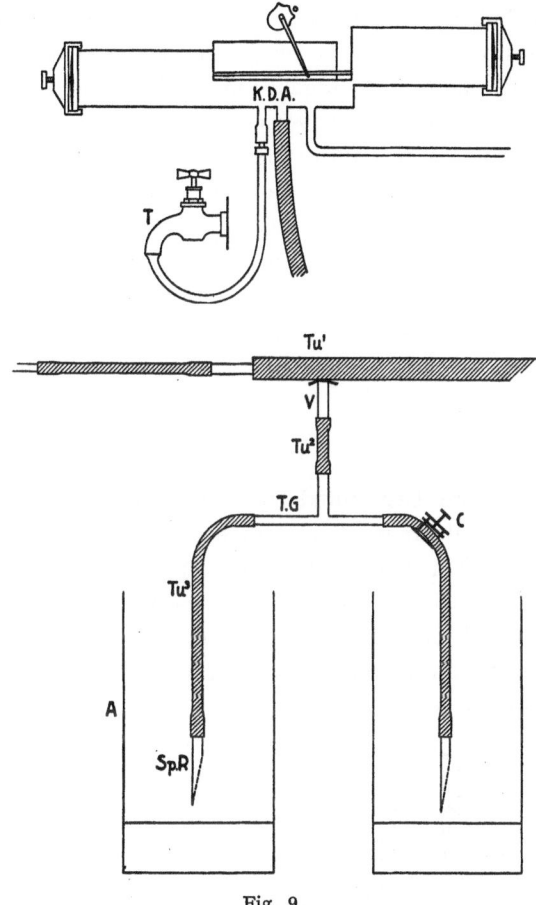

Fig. 9.

and tear, for all its movements are extremely slow, even when so many as 50 aquaria are connected to it.

It presses the air through a thin leaden tube into a thick rubber tube (Tu1) 2.5 cm in diameter and 1.5 meters long. From this tube, which was suitably studded horizontally precisely above and between two rows of aquaria, outlet was provided by punching into it small holes at suitable distances, into which was screwed the outer copper part of a bicycle valve, (V). V was connected by means of a thin rubber tube (Tu2) to a ⊥ shaped glass tube (TG) 6 mm in diameter. From each of the remaining two outlets of TG a rubber tube (Tu3) 6 mm in diameter passes into the aquarium A nearly to reach the bottom. Into the outlet of Tu3 a 4—5 cm internodal piece of spanish reed 6 mm in diameter, was pressed for 1 cm of its length. The remaining part of the reed (Sp R) was cut off wedgelike in which manner the air current was forced through the xylem vessels into the water in the form of very small bubbles. It was always necessary to adjust the pressure by means of a screw clamp (C), for naturally the pieces of spanish reed differed from aquarium to aquarium so that as a matter of course all the pieces of spanish reed did not allow exactly the same passage of air.

It was not advisable to supply air artificially when there was sufficient sunshine for then the plants supplied oxygen more efficiently.

## § 2. *Temperature*

The temperature best suited for *Lebistes* is 25°C. It is necessary to prevent great fluctuations in temperature for otherwise the fish may catch cold. This may lead to the secretion of a layer of mucus over the body, which condition on its turn offers a favourable opportunity for growth of *Saprolegnia*, a most dangerous enemy of fish. In order to maintain a constant temperature of 25° C the following method was resorted to.

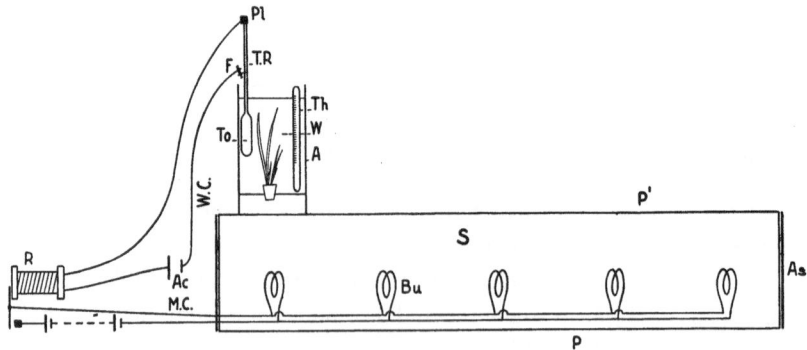

Fig. 10.

In the above figure, S represents a stove, (1 M × 0.5 M × 0.2 M) made of galvanised iron. To its vertical sides are screwed asbestos plates (As). In the inside the bottom iron plate (P) is provided with fittings for two rows of electric bulbs (Bu) of 16 candle power each, the main electric current (MC) being obtained in the ordinary manner from the main building. But the main current must be automatically switched off when the temperature of 25° C. is obtained. This is effected in the following manner: a glass tube (TR) 30 cm long and 0.5 cm in diameter, suitably studded in an aquarium (A) on the stove, is bulged out at its lower end. This lower bulged-out part is filled with clean toleol (To). The tube (TR) is then filled with clean mercury to about 7 cm from its open upper end. Sideways in the tube about mid-way of the mercury column a piece of copper wire provided with a little rod (F) with holes and screws is melted into the tube TR and reaches the mercury at a point which always remains below the level of the mercury and above the level of the toleol. The open end of the tube TR is provided with an adjustable fitting with a thin platinum wire at Pl 6 cm in length which is pressed down into the tube TR at Pl, just to reach the level of the mercury at 25° C. Fitting F is connected with an accumulator (Ac), the latter with a relais (R) which on its turn is connected with the platinum wire at Pl. As soon therefore as the platinum wire touches the level of the mercury a circuit is obtained via the relais R; through R also passes the main current MC, which is now interrupted by the relais, with the consequence that the stove starts to cool down as well as aquarium A. When the temperature has fallen something just below 25° C. the weak current (WC) from the accumulator Ac, is interrupted, the relais released, and the main current switched on again. In this manner temperature oscillations may be prevented. It is necessary to test the accumulators regularly. In the thermal regulator jar, (the aquarium A) a thermometer (Th) hangs down in the water which makes it easy to watch the temperature. The water in A should be kept approximately at the same level, and when water is to be added it must be of the same temperature as that in A.

### § 3. *Illumination*

In order to prevent decay among the aquarium plants during the shorter days of the winter season, artificial illumination was resorted to. At the suitable distances from the different shelves for the stoves in the room, bulbs (500 Watt) were fitted, the distances being determined by the simple method of counting the bubbles given off by a *Myriophyllum* twig in one of the aquaria on a cloudy afternoon without artificial illumination. This had to serve as a control experiment, for artificial illumination which agreed with that of a less cloudy day caused to great a rise of the room temperature. The room was then darkened, the lights switched on, and that distance determined where the number of bubbles per unit of time given off by the *Myriophyllum* twig roughly agreed with that obtained for the control experiment. In this manner it was possible to keep the plants growing, but $CO_2$ assimilation was very low and when the sky remained clouded for several days the buds of the plants began to show a pale, sometimes almost white colour. Generally however, artificial illumination was of great service. But these tropical fish can be bred with much less cost and trouble in tropical and subtropical countries where sufficient sunshine makes artificial illumination unnecessary. I hope to continue the *Lebistes* research in South Africa.

§ 4.  *Feeding*

This is a very important factor and it should receive special attention. I used living *Daphnias* and *Enchytreae* for the more full-grown fish, and small quantities of Wavil powder and very finely cut *Enchytreae* for young fish. Living *Enchytreae* should not be given to young fish. I have observed several cases of suffocation among young fish trying to swallow an *Enchytrea*. Finely cut *Enchytreae* may be served to young from about the 20th day after birth. Before that age they should be fed by Wavil powder served in small quantities, which gradually increase till the 20th day. Unconsumed food must be sucked out before evening sets in, for then the fish become quiet and lie on the bottom of the aquarium. It follows that the fish should never be fed when twilight is about to set in, for they should have sufficient time to consume the food. Unconsumed food is liable to decay during a long night. Feeding should take place at least twice daily, e.g. at 9 a.m. and at 3.30 p.m. For the culture of *Enchytreae* the method prescribed by WINGE (135) was followed, it is therefore unnecessary to give a detail description of the *Enchytreae* culture methods. But we may add that the substitution of bread- and bean-porridge for oatmeal-porridge gave even better results. Further experiments by myself in this direction have shown that several culture media may be used, provided that they are not sticky, as for instance is the case with agar and banana. In this medium the *Enchytrea* stretches its foremost segments, while the thicker hinder segments fail to make any progress. The little worm consequently becomes constricted in the middle and thereupon dies.

*Daphnias* were obtained from a pool outside the town but the provision lasted only till about the beginning of December. It is better to breed *Daphnias* under controlled conditions for although they are valuable food yet when obtained directly from the pool the possibility exists that unwanted organisms are caught together with the *Daphnias*, especially small *Coelenterates* as for instance the *Hydra*, which multiplies rapidly, and may in a few days time colonise the whole aquarium. This is a most dangerous situation for *Hydras* may easily be transferred to a next aquarium. They flourish under conditions necessary for the fish, some of them becoming so thick and strong that a small *Lebistes* born in such an aquarium may become entangled in the tentacles of a *Hydra*. It was very difficult to eradicate this evil. Therefore, where *Daphnias* are not kept under controlled cultural conditions, living *Enchytreae* should be preferred. However, when the *Daphnias* were used they were always washed through a net, and transferred to a basin of fresh water from which they were served. Dried *Daphnias* should be avoided as far as possible for the feeding of fullgrown *Lebistes*. But when dried *Daphnia* is to be used for young fish, it must be pulverised, and shaken in a little stoppered jar containing fresh water from which it is to be served.

We are indebted to Miss VAN HERWERDEN M.D. for her valuable hints in connection with the culture of *Daphnia*.

§ 5. *Diseases and various other abnormalities*

Experimental studies on fish demands a general knowledge of the most common diseases among fish and their treatment. It is desirable to study HOFER's "Handbuch der Fischkrankheiten" or a smaller book e.g. ROTH's "Die Krankheiten der Aquarienfische und ihre Bekämpfung". It is not my intention to give a survey of these books but only to emphasize the most common diseases met with in this connection, so that an inexperienced worker may not underrate this factor in the technique. Of the many diseases the most common are: diseases of the skin, (e.g. *Dermatomykosis saprolegniaca, Ichthyophthiriasis, Cyclochaetiasis, Gyrodactyliasis, Diplozoon paradoxum* NORDM., *Piscicoliasis, Glochidium parasiticum,* etc.), diseases of the stomach and stomach cavity (constipation, loss of appetite, *Enteritis, Ligula simplicissima*); diseases of the swimming bladder (parasites, absence of the bladder, failing of function); of the genital organs (abortus, degeneration of the eggs); of the eye; of the thyroid gland; several infectious diseases (*Tuberculosis, Lepidorthosis contagiosa, Myxoboliasis tuberosa*).

These diseases are accompanied by well defined symptoms but also by other general symptoms which may be easily recognised with a little experience, e.g. pale colour of the body, slow movements of the tail while the fish remains otherwise stationary, loss of appetite, remaining on the bottom of the aquarium, raised-up scales, etc. It is clear that the indisposition displays itself also in the general psychic behaviour of the fish.

As regards treatment, several disinfectants of weak concentration are recommended in which the fish should be bathed, e.g. table salt, ammonia, formaldehyde, kalium hypermanganicum, etc. The therapy may be studied from the books above mentioned.

Apart from these diseases I have observed several other abnormalities of which the most important cases may be described. Among the fairly well defined disturbances were the following: psychical disturbances e.g. failing to give birth to young as a result of weak sense responsiveness to external stimuli. Any harmless but abrupt change in the environmental conditions e.g. transfer to another aquarium or change in temperature will stimulate the female to give birth. Reflex disturbances, as a result of fright for instance, may be of such degree that the fish may die soon after the stimulus.

Among the blastopthorical defects were observed the following cases: e.g. what I would like to call crab eye, i.e. the eye bulges out and is very liable to get wounded; mouth defects e.g. microstomy and what may be called Habsburg lip, i.e. the lower lip projects so far that the mouth cannot be closed. Such fish have great difficulty in swollowing food and were consequently at a great disadvantage as far as development is concerned. I think there is reason to believe that this Habsburg lip is a hereditary defect, for among the young of the 2nd batch of culture 46, I observed 3 such females. The biggest of the three was crossed with her father but she died during pregnancy. Of the other embryological disturbances I may mention that of the spinal cord e.g. *Kyphosis* and *Lordosis*.

As a seperate class of disturbances may be mentioned those observed in connection with pregnancy and birth. I have observed three cases of death due to failure to give birth. These females were too young and small when crossed. Subsequent examination showed that the batches were relatively too large, for instance one of these small females contained 14 young. It might happen that a female gave birth to young, but remained indisposed and consequently died. It is a good rule not to cross females of special importance, when they are not yet fullgrown.

From certain cases observed it seemed to me that there might possibly be a genetical basis for the percentage of aborti. For most of the cultures I very seldom met with cases of aborti, but there were a few cases which regularly showed a high percentage. From culture 100 I regularly obtained from 20—25 percent cases of aborti.

From one of my $X_o Y_{El}$ cultures I could not obtain young notwithstanding the fact that the female showed the signs I usually observed when time for giving birth was due. She remained indisposed for a few days but thereafter she recovered again. During this period she had an abnormally big abdomen and would not take any food. I then noticed what seemed to me to be tiny drops of oil floating on the surface of the water. A month afterwards she showed the same signs but this time I noticed more than 20 perhaps unfertilised eggs on the bottom of the aquarium with their light vegetative poles upwards. These eggs, one after another gradually rose to the surface of the water where they appeared like oil drops. The premature loss of these eggs might be due to a relatively greater size of the eggs.

Some females showed abnormal behaviour after having given birth to young. They chased the young and might even swallow it. In such cases I made it a rule to leave the male parent with the female. She was then constantly chased by her male companion and was thus prevented to chase her young. At suitable intervals the young *Lebistes* were fished out by means of a small net and transferred to an aquarium of the same temperature as that in which they were born.

This part of the study is very interesting but it leads us too far from our object. Above all it should be emphasized that it is better to take the necessary precautions and so to prevent these diseases. The most important precautionary measures may be briefly stated as follows.

Keep instruments and other utensils scrupulously clean, always sterilize by dipping them into alcohol.

Watch the aquaria for possible growth of moulds on the lid or on the inside of the walls above the level of the water; also the rubber tube for the air supply with the piece of spanish reed is especially subject to this objection. Moulds should be regularly rinsed off by means of a piece of cotton material soaked in alcohol. Take care to prevent possible decay by removing waste material regularly. The freshness of the aquarium can be controlled by smelling at it, for in this manner the slightest trace of stagnancy of the water can be detected. Floating plants, as for instance *Myriophyllum* should be regularly rinsed in fresh water till it shows a fresh green colour. It should be noted that clear water is not always a sufficiently good sign of the fresh condition of the aquaria. Aquarium water may be kept in use as long as it smells fresh.

If possible, spend at least a few hours with the aquaria every day, preferably in the forenoon and one hour in the late afternoon in order to control the temperature for the night.

Since the *Enchytreae* are kept in culture boxes containing moist earth and other decaying matter, it is necessary to rinse the *Enchytreae*, transferring them from the one basin of fresh water to another with a clean pincette untill no dirt whatever adheres.

By strictly keeping to these rules we succeeded in breeding more than 2000 *Lebistes* without encoutering serious cases of disease. Furthermore, the better they are treated the bigger will the batches be. In connection with our *Lebistes* experiments we have even succeeded in obtaining, e.g. in the case of female E30, 59 young for her first batch, which compares very favourably with results obtained by amateurs. Generally however our fish yielded from 40—50 young per batch.

## CHAPTER II

EXPERIMENTS ON CROSSING-OVER OF THE ELONGATUS (El) AND OF THE
MACULATUS (Ma) FACTOR BETWEEN THE X- AND THE Y-CHROMOSOME
OF LEBISTES RETICULATUS

§ 1.  *The development of the colour pattern and of the elongated caudal*
*fin in the male Lebistes*

Before discussing the experiments I may briefly deal with the development of the colour pattern and of the elongation of the caudal fin.

The development of *Lebistes* is greatly retarded under less favourable conditions e.g. during the winter season, when sunshine is very scarce. Under such conditions a fish may still be immature at the age of 5 months. This may seriously complicate differentiation between round and elongated caudal fin. It was therefore desirable to determine the stage of development of the elongation of the caudal fin. I studied several such cases, the following of which may serve as an illustration. See Fig. 11.

Male 32, born July 14th, 1931.

22/ 9/31, 1.  Anal fin commences to metamorphose; dorsal fin slightly pigmented; caudal fin round; fish approximately 3/4 of fullgrown size.

24/ 9/31, 2.  Gonopodium nearly completely metamorphosed; dark black spot in dorsal fin.

26/ 9/31, 3.  Anal fin completely metamorphosed into gonopodium.

28/ 9/31, 4.  Mother-of-pearl shining on the side of the body behind gill-cleft.

1/10/31, 5.  Vague orange colouring on side of body below dorsal fin.

3/10/31, 6.  Deep orange colouring on side of body below dorsal fin; dark longitudinal shading near anus; vague blue shade near root of caudal fin and end of tail; a vague greenish shade on side of body behind gill-cleft; perhaps faint sulphur streak on upper edge of caudal fin.

5/10/31, 7.  Faint sulphur shade on upper edge of caudal fin; perhaps very faint sulphur shade on lower edge of caudal fin.

8/10/31, 8.  Black longitudinal patch near anus much darker; slight orange colouring on upper edge of caudal fin; a narrow sulphur streak on lower edge of caudal fin, but shorter than shading on upper edge.

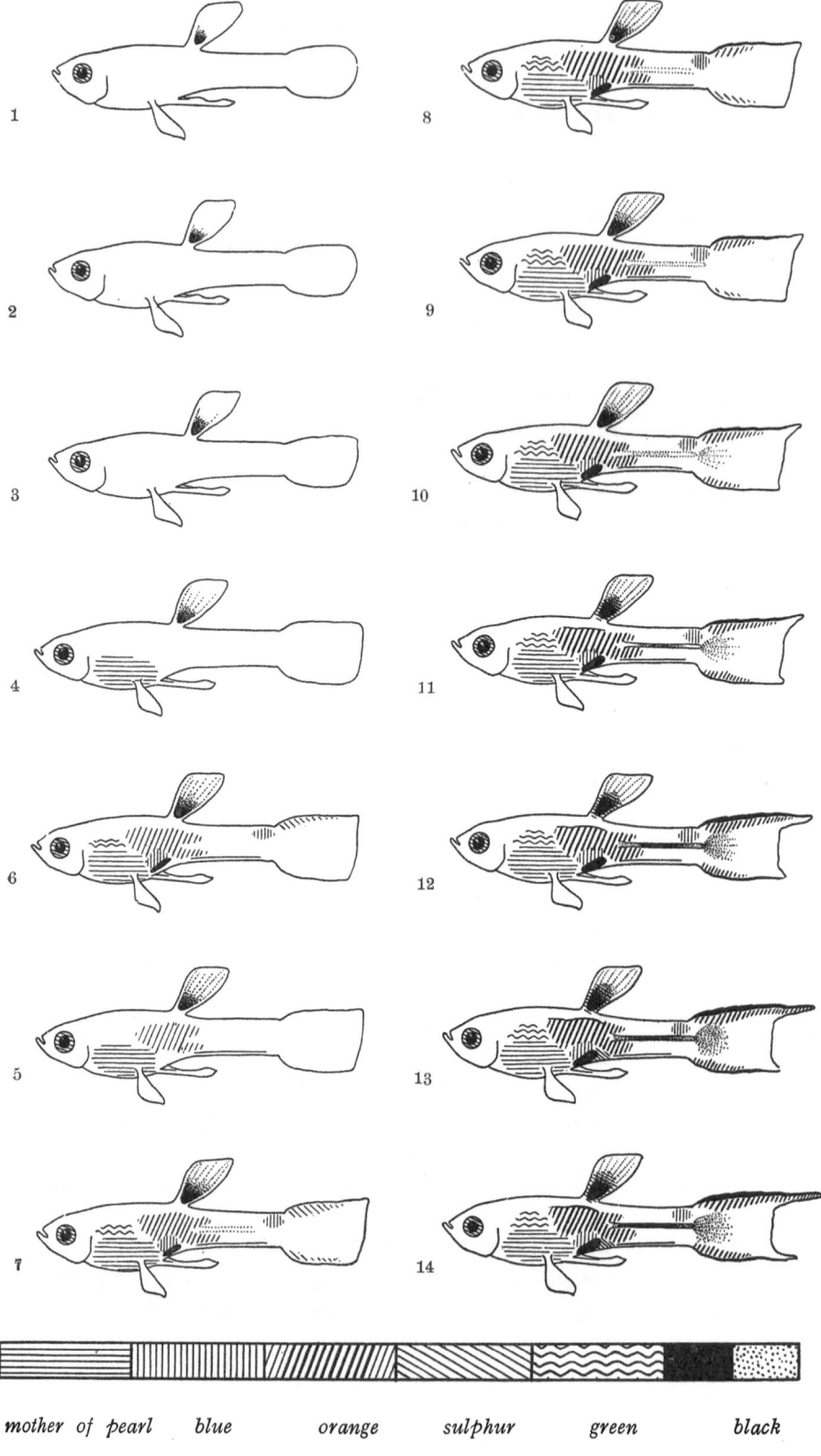

mother of pearl     blue        orange      sulphur      green        black

Fig. 11.

10/10/31,  9. Deeper orange colouring on upper edge of caudal fin; slight orange streak on lower edge of caudal fin.

11/10/31, 10. A fairly distinct dark seam along upper edge of caudal fin; upper edge of caudal fin distinctly pointed.

14/10/31, 11. Sulphur colouring around black spot in dorsal fin; caudal fin slightly more elongated and more distinctly pointed at upper end; slight crescent-like shading below dorsal fin passing over orange patch below dorsal fin.

18/10/31, 12. Blue tinge around anal spot; a fair amount of melanin pigmentation in middle region of caudal fin; both upper and lower edges of caudal fin distinctly pointed, especially the upper edge.

22/10/31, 13. A distinct sulphur shade behind anal spot; fish full-grown with brilliant colour pattern and elongated caudal fin.

14. Complete stage of a more than one year old fish.

From these observations it is apparent that the factor for elongation of the caudal fin expresses itself relatively later than most of the other factors observed. Several other cases were studied giving the same general result. In the case of No. 46 almost all the colours were present and brilliant on 21/12/31, but not untill 30/12/31 was elongation of the caudal fin observed. It may even be possible that elongation of the caudal fin takes place after maturity. These facts must be duly considered when the fish are to be differentiated.

§ 2. *Crossing-over of the elongatus factor.*

In June 1917 WINGE mated his "Poppelgade" male *Lebistes* $X_{El}Y_{Ma}$ to a virgin female of the constitution $X_oX_o$. The phenotype of the male was maculatus (Ma), a gene either working pleiotropically on, or absolutely linked to the factors for (a) great black spot on dorsal fin, (b) great red side patch beneath and behind dorsal fin, (c) black spot near anus. This gene for maculatus was found to be absolutely linked to the Y-chromosome and perhaps it may be more correct to say that the chromosome is Y, when maculatus is present in it. The factors so far are those of the ordinary $X_oY_{Ma}$ male; but the "Poppelgade" male was an exceptional male possessing a caudal fin with elongated upper edge, besides it was of a yellowish red colour as far as the upper and lower edges were concerned. This pleiotropic factor was named elongatus (El). From the above mentioned cross, $X_oX_o \times X_{El}Y_{Ma}$ 74 offspring were reared, 73 of which had no elongated fin, the remaining one was elongatus. It resembled a case of criss-cross inheritance and a case of crossing-over. Consequently 'El' according to WINGE was sex-linked in contradistinction to 'Ma' which was sex-limited. The diagram is as follows:

$$X_oX_o \quad \times \quad X_{El}Y_{Ma}$$

offspring, $X_oX_{El}$ ♀♀, $X_oY_{Ma}$ ♂♂, non crossovers: round caudal fin;

$$X_oX_o \quad \times \quad X_oY_{Ma,El}$$

offspring, $X_oX_o$ ♀♀, $X_oY_{Ma,El}$♂♂, crossovers: elongated caudal fin. If so then the crossing $X_oY_{Ma}$ males $\times$ $X_oX_{El}$ females should produce $X_oX_o, X_oX_{El}$ ♀♀; $X_oY_{Ma}, X_{El}Y_{Ma}$ ♂♂ in equal proportions, which were actually found: 26 $X_{El}Y_{Ma}$ and 27 $X_oY_{Ma}$ ♂♂. The $F_2$ females should be $X_oX_o$ and $X_{El}X_o$ in the ratio 1 : 1 which was also verified. Furthermore WINGE obtained 64 elongatus males and 4 non-elongatus males from the cross $X_oX_o \times X_oY_{Ma,El}$. The 4 non-elongatus males were derived from the maternal gamete $X_o$ and a paternal cross-over gamete $Y_{Ma}$, resulting in the zygote $X_oY_{Ma}$. So that according to WINGE's data: the factor El was transferred from X to Y, 1 out of 74 cases; from Y to X, 4 out of 68 cases. The difference in frequency seemed striking, in the case from X to Y, 1 : 74, in that of from Y to X, 1 : 17, the respective c.o.v.'s being approximately 5.9 % and 1.35 %. WINGE says: "It is remarkable that crossing-over has not been equally frequent in both directions; however, as the amount of material was not very large I do not feel justified in attaching any significance to this fact".

The difference in frequency of crossing-over of El between the cases where it is carried by the X- and where it is carried by the Y-chromosome involves an important cytological problem namely what may be called the exchange adaptation of the recipient to the donor chromosome. Since the problem was not fully made out by WINGE, we decided to take up the matter further.

On the 25th May 1931 we received from Prof. WINGE four aquaria, N° 1 containing 10 $X_oY_{Ma,El}$ males, N° 2 containing 15 $X_oY_{Ma}$ and $X_{El}Y_{Ma}$ types, N° 3 containing 50 fullgrown $X_oX_o$ virgin females, and N° 4 containing $F_1$ offspring from the cross $X_oX_o \times X_oY_{Ma}$.

In order to determine the frequency of transfer of El from Y to X the following type of cross was made use of: $X_o X_o \times X_o Y_{Ma,El}$ from which were to be expected $X_o X_o$ females and $X_o Y_{Ma,El}$ males as non-crossovers, and $X_o X_{El}$ females and $X_o Y_{Ma}$ males as crossovers.

To determine the frequency of transfer of El from X to Y the following type of cross was made: $X_o X_o \times X_{El} Y_{Ma}$ from which were to be expected $X_o X_{El}$ females and $X_o Y_{Ma}$ males as non-crossovers. This type of cross presented a special case of criss-cross inheritance, the $F_1$ non-crossover males resembling their female parent, the $F_1$ non-crossover females resembling their male parent, but for the fact that the El factor does not as a rule display itself in the female.

There are certain secondary differences between the two above mentioned types of crosses. From the cross $X_o X_o \times X_o Y_{Ma,El}$, males of the constitution $X_o Y_{Ma,El}$ are easily obtained. They may also be obtained from the crossover gametes of the cross $X_o X_o \times X_{El} Y_{Ma}$. Besides, these $X_o Y_{Ma,El}$ males are easily tested out. It is however, much more difficult from the cross $X_o X_o \times X_{El} Y_{Ma}$ to obtain $X_{El} Y_{Ma}$ males, for they are to be obtained from the $F_2$, that is to say by crossing the $F_1$ $X_o X_{El}$ females with $X_o Y_{Ma}$ males, from which are to be expected $X_o Y_{Ma}$ and $X_{El} Y_{Ma}$ males in equal proportions. And since the El factor does not show itself in the female, these $F_1$ $X_o X_{El}$ females have to be tested out by suitable crosses.

The facts so far obtained may be summarised as follows, the last column showing the total number of offspring obtained the greater part of which is not yet differentiated:

TABLE 6.   $X_o X_o \times X_o Y_{Ma,El}$

| cross | ♀♀ | non-cross-over ($X_o$-$Y_{Ma,El}$)♂♂ | crossover ($X_o Y_{Ma}$) ♂♂ | ♀♀ + ♂♂ | offspring obtained |
|---|---|---|---|---|---|
| 31 | 26 | 28 | | 54 | 57 |
| 40 | 39 | 45 | | 84 | 129 |
| 41 | 16 | 13 | | 29 | 162 |
| 42 | 30 | 33 | | 63 | 217 |
| 45 | 34 | 26 | | 60 | 238 |
| 46 | 34 | 27 | 3 | 64 | 120 |
| 51 | 2 | 7 | | 9 | 96 |
| total | 181 | 179 | 3 | 363 | 1019 |
| later additions | | 93 | 2. | | |
| grand total | | 272 | 5 | | |

From this table it is apparent that the only previous crossovers observed were the 3 $F_1$ males from cross 46. This suggests a genetical basis for the high frequency of crossing-over in the male parent 46, its c.o.v. amounting to 10 percent. In the hope to obtain more certainty as to a possible genetical basis and in this manner to select a race with a relatively high frequency of crossing-over, brother-sister crosses were made with the $F_1$'s from cross 46. From one of these crosses we have so far obtained 143 offspring, but owing to the intervening winter months they are not as yet full-grown. The experiment will be continued.

The results so far obtained from the crosses of the type $X_o X_o \times X_{El} Y_{Ma}$ may be briefly summarised in Table 7.

TABLE 7.   $X_o X_o \times X_{El} Y_{Ma}$

| cross | ♀♀ | El ♂♂ | non-El ♂♂ | ♀♀ + ♂♂ | offspring obtained |
|---|---|---|---|---|---|
| 24 | | | | | 15 |
| 30 | 1 | | | 1 | 99 |
| 37 | 25 | 28 | 1 | 54 | 114 |
| 47 | 7 | 4 | | 11 | 86 |
| 41 | | | | | 71 |
| 50 | 12 | 6 | | 18 | 26 |
| 100 | | | | | 33 |
| total | 45 | 38 | 1 | 84 | 444 |
| later additions | | 52 | 2 | | |
| grand total | | 90 | 3 | | |

It took a long time before offspring were obtained from these $X_o X_o \times X_{El} Y_{Ma}$ crosses, though the treatment was the same as that of the other type of crosses. From some of these $X_o X_o \times X_{El} Y_{Ma}$ crosses, not included in the above summary, no offspring were obtained. However, the results so far obtained from the successful crosses are sufficient to show that they are contrary to all expectations, for as said above, the majority of male offspring were expected to be of the non-El type (round caudal fin) and only the crossovers were expected to be of the El type. As appears from the above summary only 3 males with a round caudal fin were observed among the offspring from the $X_o X_o \times X_{El} Y_{Ma}$ crosses, while 90 male offspring had an elongated caudal fin. As soon as this fact became apparent to me

I communicated the first results to Prof. WINGE who answered me giving his opinion about the matter. He also thinks it a peculiar thing that the only supposed $X_{El}Y_{Ma}$ type which has given young at that time, gave 3 males all having an elongated caudal fin. The supposed $X_{El}Y_{Ma}$ parent according to Prof. WINGE, was himself obtained from a cross $X_{El}X_{El} \times X_{El}Y_{Ma}$ and WINGE thinks that this young was derived from a crossover gamete $Y_{Ma,El}$ from his father $X_{El}Y_{Ma}$, so that this young actually was of the constitution $X_{El}Y_{Ma,El}$. This explanation would have been a very plausible one but for the fact that this unexpected result was observed in connection with all my supposed $X_{El}Y_{Ma}$ males. Furthermore, the presence of one young with a round caudal fin (cross 37) complicates the matter, for in no wise may it be expected from the cross $X_{El}X_{El} \times X_{El}Y_{Ma,El}$. This male was reobserved more than two months later but still showed a round caudal fin. Lateron 2 additional non-El $\male\male$ were observed.

In the hope of solving the problem, $F_1$ females derived from cross 37 ($X_oX_o \times X_{El}Y_{Ma}$) were crossed to $X_oY_{Ma}$ males. In case the parent was of the consitution $X_{El}Y_{Ma,El}$ as WINGE suggested, we should expect the $X_oX_{El}$ females and $X_oY_{Ma,El}$ males in equal proportions. These $F_1$ males and females may now be tested further. The $F_1$ females when crossed to $X_oY_{Ma}$ males will be expected to give apart from the female offspring, $X_oY_{Ma}$ and $X_{El}Y_{Ma}$ males in equal proportions. From this type of cross we have obtained 100 young more or less, and we hope to differentiate them within a month or two. The results so far obtained suggest that the males were of the formula $X_oY_{Ma,El}$, as a consequence of which the $F_1$ males from the supposed $X_{El}Y_{Ma}$ males are also being tested out. All $\male\male$ from the test-crosses are non-El.

These complications in connection with the crosses of the type $X_oX_o \times X_{El}Y_{Ma}$ makes it as yet impossible to compare the frequencies of transfer of the El factor from X to Y and from Y to X. The possible hereditary basis of the difference in crossing-over among the crosses of the type $X_oX_o \times X_oY_{Ma,El}$ offers another difficulty. If we treat the case of cross 46 purely as a matter of chance then we may perhaps say that according to our results the El factor transits from Y to X in 5 cases out 277 of the giving a c.o.v. of 1.8 percent. It will be remembered that WINGE observed transition of the El factor from Y to X in 4 cases out of 68 giving a c.o.v. of 5.9 percent. If the c.o.v. is calculated on WINGE's material and my own, it will amount to

2.6 percent which is a slightly higher percentage than that obtained by Winge in the case of transition of El from X to Y which occurred once in 74 cases or 1.35 percent of the cases, but it is a serious statistical mistake to calculate crossing-over when only one case was observed, for not untill at least a second case is observed may we form any notion as to the denominator of the crossing-over fraction. This however means that there is chance for enlargement of the denominator, that is for a decrease in the c.o.v. It is probable therefore that the El factor crosses over from Y to X somewhat more frequently than from X to Y.

### § 3. *Crossing-over of the maculatus factor*

One of the most important results obtained by Winge in connection with his *Lebistes* experiments is the phenomenon of sex-limited inheritance (Y-linked), which up till that time was quite unknown. Among others the gene for maculatus behaved in this sex-limited manner, namely, it was transmitted from father to all sons, generation after generation, but was never in any circumstances inherited by the female offspring (Winge, 135). Thousands of specimens were investigated by Winge without a single case having been observed where this gene crossed over from the Y- to the X-chromosome. It should therefore either be situated so close to the male determining gene in the Y-chromosome that crossing-over becomes practically (physically) impossible, or the characters maculatus and maleness should be the result of a pleiotropic action of one and the same gene. I cannot agree with Winge's alternative: "or it must be identical with that (male determining) gene which is to say that there must be multiple allelomorphism in regard to the male determining factor". However, Winge acknowledges the fact that crossing-over of the maculatus factor may nevertheless be possible so that this problem must be left to be decided by future experiments. We may briefly state that we have possibly found the answer: the maculatus factor may cross over from X to Y, after all. Among our cultures we kept the stock $X_oX_o \times X_oY_{Ma}$ from which cross were obtained several hundreds of $X_oX_o$ females and $X_oY_{Ma}$ males. However we were surprised to observe among them a female with a distinct black spot in her dorsal fin, while she was a female in all other respects. Three months have elapsed since and she does not show any sign of maleness. She is full-grown, being about $5\frac{1}{2}$ months old and is crossed to a $X_o Y_{Ma}$ male. Since she is supposed to be a $X_o X_{Ma}$ female by crossing-over from Y to X, it is to be expected that her offspring will be $X_o X_{Ma}$ and $X_o X_o$ females in the ratio 1 : 1 as well as $X_{Ma} Y_{Ma}$ and $X_o Y_{Ma}$ males in the ratio 1 : 1.

Apart from my own cultures I have received a similar female from Mr. MOORLAG, the president of a local aquarium society. This female was noticed among some 300 *Lebistes* young derived from the cross of an ordinary *Lebistes* female with a male having a round caudal fin. The male was obtained because of its bright colouring, but was not known to have any other exceptional characters. From this cross some 300 young were obtained among which the female with a black spot in the dorsal fin was observed and intentionally cared for, being also crossed to an ordinary *Lebistes* male, and she has since given birth to six young, all males.

From these two females I hope to cultivate a $X_{Ma}$ race which may perhaps be of special interest to the quantitative theory of sex which of lately has been advanced by WINGE (136) also for *Lebistes*.

These *Lebistes* experiments are to be interrupted for the time being, owing to my departure from Groningen for South Africa, where I hope to continue them. As soon as further results are obtained especially in connection with crossing-over of the maculatus factor they are to be communicated to Prof. TAMMES and Prof. WINGE.

## CHAPTER III

THE EFFECT OF CENTRIFUGATION AND OF ULTRA VIOLET LIGHT ON CROSSING-OVER BETWEEN BLACK AND VESTIGIAL IN THE SECOND CHROMOSOME OF DROSOPHILA MELANOGASTER, WITH A COMPARISON OF THE RESULTS FROM REPULSION AND FROM COUPLING BACK-CROSSES

This region of the 2nd chromosome was chosen intentionally, because several investigations have shown that the central region is very sensitive to changes in crossing-over due to some external physical agents, e.g. temperature and X-rays. Furthermore the region is not too short, crossing-over taking place fairly frequently. It is however too short to allow for frequent double crossing-over. The phenotypes are easily differentiated and the viability is satisfactory.

From the onset of the experiments the results obtained from the repulsion back-crosses were treated apart from the results obtained from the coupling back-crosses. In the former case the $F_1$ females were derived from the cross black $\times$ vestigial, in the latter case from the cross Florida wild $\times$ black purple (curved) vestigial arc speck. The stocks were obtained from Dr. CURT STERN by Miss M. A. VAN HERWERDEN M.D. To both Dr. VAN HERWERDEN and Dr. STERN we are greatly indebted. The controls as well as the treated were constantly kept at 22° C. and transferred to fresh jars every 5 or 6 days. Except for the treatment in question the conditions were kept constant as far as possible. In all cases the $F_1$ females were back-crossed with a black purple (curved) vestigial arc speck male.

## § 1. *Centrifugation*

For the centrifugation experiments the velocity was 2000 revolutions per minute at a distance of 14 cm from the axle of the centrifuge. The duration of the treatment was varied.

The results may be summarised in the following table, the classes being always given in the order black long (bV); gray vestigial (Bv), gray long (BV), black vestigial (bv), the black long and gray vestigial classes being the crossover classes for the coupling back-crosses while the gray long and black vestigial classes are crossovers in the case of repulsion back-crosses. In the text b.c. = back-cross; M c.o.v. = mean crossover value percent; $m = \dfrac{\sigma}{\sqrt{n}}$; diff. $M_1, M_2,$ = difference between two mean values $M_1$, and $M_2$; m diff. $= \pm \sqrt{m^2_1 + m^2_2}$; 1st bottles,

TABLE 8. Repulsion b.c. Offspring from $F_1$ females which underwent centrifugation.

| | 1st bottles | | | | 2nd bottles | | | | 3rd bottles | | | |
|---|---|---|---|---|---|---|---|---|---|---|---|---|
| culture | bV | Bv | BV | bv | bV | Bv | BV | bv | bV | Bv | BV | bv |
| 99 | 89 | 92 | 41 | 28 | 52 | 58 | 14 | 13 | | | | |
| 100 | 81 | 96 | 43 | 41 | 74 | 57 | 9 | 14 | 31 | 19 | 5 | 3 |
| 131 | 50 | 55 | 12 | 18 | 43 | 40 | 8 | 9 | | | | |
| 132 | 51 | 53 | 8 | 16 | 32 | 51 | 4 | 10 | 62 | 50 | 8 | 8 |
| total | 271 | 296 | 104 | 103 | 201 | 206 | 35 | 46 | 93 | 69 | 13 | 11 |
| c.o.v. | 207/774 = 26.74% | | | | 81/488 = 16.59 % | | | | 24/186 = 12.9 % | | | |
| M c.o.v. | 25.07, m ± 2.5 | | | | 16.77, m ± 0.96 | | | | 13.15 m, ± 0.46 | | | |

2nd bottles, etc., = offspring obtained during 1st period of six days, 2nd period of six days, etc.

TABLE 9. Repulsion b.c. Offspring from $F_1$ control females.

| culture | 1st bottles | | | | 2nd bottles | | | | 3rd bottles | | | |
|---|---|---|---|---|---|---|---|---|---|---|---|---|
| | bV | Bv | BV | bv | bV | Bv | BV | bv | bV | Bv | BV | bv |
| 103 | 131 | 121 | 33 | 34 | 58 | 51 | 11 | 16 | | | | |
| 135 | 75 | 70 | 18 | 18 | 45 | 44 | 8 | 8 | 52 | 48 | 8 | 8 |
| total | 206 | 191 | 51 | 52 | 103 | 95 | 19 | 24 | 52 | 48 | 8 | 8 |
| c.o.v. | 103/500 = 20.6 % | | | | 43/241 = 17.84 % | | | | 16/116 = 13.8 % | | | |
| M c.o.v. | 20.44, m ± 0.39 | | | | 17.54, m ± 1.6 | | | | | | | |

The difference between the controls and the treated for the first period of 6 days was 4.63 units of crossing-over, m diff. being ± 2.53, a small but perhaps significant difference in this connection, for, 3 out of the 4 experiments, viz. 99, 100 and 131 show higher c.o.v.'s than the controls. For the second period of 6 days no such difference of any significance was observed between the treated and the controls, the difference being 0.77, m ± 1.6.

The grand totals for all the repulsion experiments in connection with centrifugation were 565, 571, 152, 160 giving a very reliable c.o.v. of 21.55 %, while the grand totals for the controls were 443, 390, 97, 106 also giving a very reliable value of 19.59 %. This means that there was a difference of approximately two units of crossing-over between the treated and the controls for the repulsion experiments even if age-analysis was neglected for the first 18 days.

As regards difference in duration of treatment the following results were obtained for the repulsion experiments.

TABLE. 10. Repulsion b.c. The effect of duration of centrifugation on crossing-over between b and v.

| duration | culture | M c.o.v. 1st bottles | M c.o.v. 2nd bottles | M c.o.v. 3rd bottles |
|---|---|---|---|---|
| 1 min. | 99, 100, | 29.67, m ±1.58 | 17.82, m ±1.28 | 13.8 |
| 5 min. | 132 | 18.75 | 14.4 | 12.5 |
| 10 min. | 131 | 22.22 | 17.0 | |
| | control | 20.44, m ±0.39 | 17.54, m ±0.64 | 13.79 |

I do not think that these data on the duration of treatment show any general rule. Once it is known that centrifugation has an effect on crossing-over the question as to the effect of the duration of treatment may be specially studied in the same manner as PLOUGH studied the effect of temperature on crossing-over. Nevertheless, we may note a few interesting points. In the first instance cultures 99 and 100 although being treated for one minute only, gave the highest c.o.v. viz. 29.67 %, the individual c.o.v.'s being 27.4 % and 31.92 %. For these cultures we furthermore notice a very striking drop in the c.o.v. for the 2nd bottles, the difference being 11.85, m $\pm$ 2.1, so that it is more than 5 $\times$ m diff.

TABLE 11.   Coupling b.c. Offspring from $F_1$ females which underwent centrifugation.

| culture | 1st bottles | | | | 2nd bottles | | | | 3rd bottles | | | |
|---|---|---|---|---|---|---|---|---|---|---|---|---|
| | bV | Bv | BV | bv | bV | Bv | BV | bv | bV | Bv | BV | bv |
| 130 | 14 | 7 | 76 | 45 | 11 | 6 | 45 | 49 | 5 | 10 | 32 | 56 |
| 133 | 8 | 10 | 51 | 49 | 15 | 14 | 68 | 74 | 3 | 2 | 11 | 14 |
| 134 | 10 | 13 | 61 | 58 | 4 | 10 | 63 | 70 | 13 | 15 | 55 | 60 |
| 137 | 7 | 12 | 50 | 45 | 7 | 5 | 43 | 37 | 6 | 15 | 55 | 59 |
| 138 | 12 | 12 | 54 | 51 | 6 | 10 | 48 | 61 | 3 | 2 | 22 | 10 |
| 139 | 3 | 2 | 36 | 27 | 8 | 7 | 50 | 61 | | | | |
| 141 | 4 | 4 | 58 | 31 | 15 | 18 | 88 | 89 | 8 | 5 | 39 | 23 |
| 142 | 12 | 10 | 9 | 41 | 21 | 12 | 81 | 76 | 4 | 7 | 37 | 52 |
| 143 | 10 | 7 | 31 | 72 | 9 | 6 | 37 | 36 | 9 | 15 | 72 | 62 |
| total | 80 | 77 | 466 | 419 | 86 | 83 | 498 | 531 | 48 | 69 | 321 | 326 |
| c.o.v. | 157/1042=15.06% | | | | 169/1198=14.11% | | | | 117/764=15.31% | | | |
| M c.o.v. | 14.56, m $\pm$ 0.58 | | | | 14.4, m $\pm$ 0.86 | | | | 15.39, m $\pm$ 0.4 | | | |

TABLE 12.   Individual c.o.v.'s from table 11.

| culture | c.o.v. % 1st bottles | c.o.v. % 2nd bottles | c.o.v. % 3rd bottles |
|---|---|---|---|
| 130 | 14.86 | 15.31 | 12.3 |
| 133 | 15.25 | 16.95 | 16.7 |
| 134 | 16.2 | 9.52 | 19.6 |
| 137 | 16.67 | 13.04 | 15.6 |
| 138 | 18.61 | 12.8 | (13.51) |
| 139 | 7.35 | 11.9 | |
| 141 | 8.25 | 15.71 | 17.33 |
| 142 | 19.64 | 17.37 | 11.0 |
| 143 | 14.17 | 17.04 | 15.2 |
| M c.o.v. | 14.56, m $\pm$ 0.58 | 14.40, m $\pm$ 0.86 | 15.39, m $\pm$ 0.4 |

As far as the coupling experiments are concerned the results also show a slightly but possibly important, higher c.o.v. for the 1st bottles of the treated flies as compared with the 1st bottles of the controls.

We have to make some correction in these mean c.o.v.'s. By studying table 11, it becomes apparent especially for the 1st bottles, that the reliability of cultures 130, 139, 141, and 143 is weakened because of the weak representation of the subclasses of either the cross-over or the non-crossover class. The remaining cultures 133, 134, 137, 138, 142 give a mean c.o.v. of 17.27, m $\pm$ 0.5 for the 1st bottles and 13.94, m $\pm$ 1.15 for the 2nd. The totals were as follows.

TABLE 13.  Totals of well represented cultures according to table 11.

|  | 1st bottles | | | | 2nd bottles | | | |
|---|---|---|---|---|---|---|---|---|
| cultures | bV | Bv | BV | bv | bV | Bv | BV | bv |
| 133 + 134 + 137 + 138 + 142 | 49 | 57 | 265 | 244 | 53 | 51 | 303 | 318 |
| c.o.v. | 106/615 = 17.23 % | | | | 104/725 = 14.34 % | | | |
| M c.o.v. | 17.27, m $\pm$ 0.5 | | | | 13.94, m $\pm$ 1.15 | | | |
| M c.o.v. (control) | 14.47, m $\pm$ 0.45 | | | | 15.35, m $\pm$ 3.28 | | | |

From these data it is apparent that each of the well represented cultures show a significantly higher c.o.v. for the 1st bottles as compared with the 1st bottles of the controls. The difference is 2.8, while m diff. is $\pm$ 0.66, so that the difference is approximately 4 $\times$ m diff. For the second bottles the c.o.v. difference between the treated and the controls was not found to be statistically significant, the errors involved for the mean value for the second bottles make the mean c.o.v.'s less reliable criteria than the c.o.v. calculated from grand totals which were as follows for the controls:

TABLE 14.  Coupling b.c.  Control data.

| 1st bottles | | | | 2nd bottles | | | |
|---|---|---|---|---|---|---|---|
| bV | Bv | BV | bv | bV | Bv | BV | bv |
| 13 | 13 | 80 | 70 | 12 | 13 | 67 | 87 |
| c.o.v. 26/180 = 14.4 % | | | | 25/179 = 13.96 % | | | |

From these data it is apparent that as to the 2nd bottles no significant difference can be noted between the c.o.v.'s of the treated and of the controls.

As regards the effect of different durations of treatment the following figures may be given. Cases marked thus ? are not taken into account because of the weak representation of the sub-classes.

TABLE 15.    Centrifugation 5 minutes, coupling b.c.

| culture | 1st bottles | | | | 2nd bottles | | | | 3rd bottles | | | |
|---|---|---|---|---|---|---|---|---|---|---|---|---|
| | bV | Bv | BV | bv | bV | Bv | BV | bv | bV | Bv | BV | bv |
| 130 | 14 | 7 | 76 | 45? | 11 | 6 | 45 | 49? | 5 | 10 | 52 | 56 |
| 137 | 7 | 12 | 50 | 45 | 7 | 5 | 43 | 37 | 6 | 15 | 55 | 59 |
| 139 | 3 | 2 | 36 | 37 | 8 | 7 | 50 | 61 | | | | |
| total | 24 | 21 | 162 | 127 | 26 | 18 | 138 | 147 | 11 | 25 | 107 | 115 |
| total — 130 | 10 | 14 | 86 | 82 | 15 | 12 | 93 | 98 | | | | |
| c.o.v. | 24/192 = 12.5 % | | | | 27/218 = 12.4 % | | | | 36/258 = 13.95 % | | | |
| M c.o.v. | 12.96, m ± 2.2 | | | | 13.41, m ± 0.73 | | | | 13.95, m ± 1.72 | | | |

TABLE 16.    Centrifugation 10 minutes, coupling b.c.

| culture | 1st bottles | | | | 2nd bottles | | | | 3rd bottles | | | |
|---|---|---|---|---|---|---|---|---|---|---|---|---|
| | bV | Bv | BV | bv | bV | Bv | BV | bv | bV | Bv | BV | bv |
| 133 | 8 | 10 | 51 | 49 | 15 | 14 | 68 | 74 | 5 | 10 | 52 | 56 |
| 138 | 12 | 12 | 54 | 51 | 6 | 10 | 48 | 61 | | | | |
| total | 20 | 22 | 105 | 100 | 21 | 24 | 116 | 135 | 5 | 10 | 52 | 56 |
| c.o.v. | 42/247 = 17.0 % | | | | 45/296 = 15.2 % | | | | 15/123 = 12.3 % | | | |
| M c.o.v. | 16.93, m ± 1.2 | | | | 14.87, m ± 1.5 | | | | | | | |

TABLE 17.    Centrifugation 15 minutes, coupling b.c.

| culture | 1st bottles | | | | 2nd bottles | | | | 3rd bottles | | | |
|---|---|---|---|---|---|---|---|---|---|---|---|---|
| | bV | Bv | BV | bv | bV | Bv | BV | bv | bV | Bv | BV | bv |
| 134 | 10 | 13 | 61 | 58 | 4 | 10 | 63 | 70 | 13 | 15 | 55 | 60 |
| 141 | 4 | 4 | 58 | 31 | 15 | 18 | 88 | 89 | 8 | 5 | 39 | 22 |
| total | 14 | 17 | 119 | 89 | 19 | 28 | 151 | 159 | 21 | 20 | 94 | 82 |
| c.o.v. | 31/239 = 12.97% | | | | 47/357 = 13.2 % | | | | 41/217 = 18.4 % | | | |
| M c.o.v. | 12.2, m ± 2.81 | | | | 12.6, m ± 2.2 | | | | 18.5, m ± 0.8 | | | |

TABLE 18. Centrifugation 20 minutes, culture 142; and 25 minutes, culture 143. Coupling b.c.

| | 1st bottles | | | | 2nd bottles | | | | 3rd bottles | | | |
|---|---|---|---|---|---|---|---|---|---|---|---|---|
| culture | bV | Bv | BV | bv | bV | Bv | BV | bv | bV | Bv | BV | bv |
| 142 | 12 | 10 | 49 | 41 | 21 | 12 | 81 | 76 | 4 | 7 | 37 | 52 |
| 143 | 10 | 7 | 31 | 72 | 9 | 6 | 37 | 36 | 9 | 15 | 72 | 62 |
| total | 22 | 17 | 80 | 113 | 30 | 18 | 118 | 112 | 13 | 22 | 109 | 114 |
| c.o.v. | 39/232 = 16.8 % | | | | 48/287 = 16.51 % | | | | 35/258 = 13.6 % | | | |
| M c.o.v. | 16.9, m ±1.9 | | | | 17.2, m ± 0.1 | | | | 13.1, m ± 1.5 | | | |

After the necessary corrections in connection with the cases where the sub-classes for either the crossover or the non-crossover classes are not proportionally represented, we obtain the following summary:

TABLE 19. Summary of effect of duration of treatment on the % of c.o. Coupling b.c.

| min. | 1st bottles | 2nd bottles | 3rd bottles |
|---|---|---|---|
| 5 | 12.96, m ± 2.2 | 13.41, m ± 0.73 | |
| 10 | 16.93, m ± 1.2 | 16.95 | 12.3 |
| 15 | 16.2 | 15.78 | 18.4 |
| 20 | 19.64 | 17.37? | 11.0? |
| 25 | | 17.04 | 15.2 |

From this summary it appears as if there is a rough positive correlation between the duration of centrifugal treatment and crossing-over between black and vestigial in the case of coupling b.c. experiments. It must be admitted however, that the figures are not fully convincing and further experiments will be carried out in order to obtain more certainty. It will be remembered that in this connection no system was found for the repulsion back-crosses.

For the coupling experiments, also the grand total displays a higher c.o.v. for the treated females than for the controls. For the treated females the class totals which were very well represented are 166 bV, 160 Bv, 944 BV, 950 bv giving a crossing-over value of 14.68 % while the controls gave the following class totals: 32, 38, 216, 226 with a c.o.v. of 13.67 %.

Concluding the centrifugation experiments it seems justified to state that centrifugation caused a small rise in crossing-over between black and vestigial for the first period of 6 days. The effect of the duration of treatment is as yet doubtful.

## § 2. *Ultra violet light*

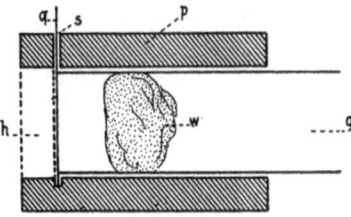

Fig. 12.

For the experiments with ultra violet light a quartzglass quicksilver lamp was made use of. The driving current was 3.5 amperes at 160 volts. The flies which were to be treated were carefully transferred (see fig. 12) to a glass tube (g) 5 cm long and 1.5 cm in diameter, the bottom of the tube was previously cut out and edges polished. Into this end of the tube a plug of wadding (w) was pressed before the flies were transferred into it. As soon as the flies were transferred the tube was quickly pressed into a hole (h) of the same dimensions bored through a piece of plank (p) $4 \times 3 \times 3$ cm. Near the opposite opening of the hole a slit (s) was sawn at right angles to it and reaching deeper than the hole itself. Into this slit a piece of quartzglass (q) $2 \times 2$ cm was pressed. The plug at the other end of the tube was carefully pressed down till there was left just enough space between the plug and the quartzglass for the flies to move about. The flies were then treated, care being given to the fact that the surface of the quartzglass should be parallel with the quartzglass tube of the burner. Distance and duration were varied for the different experiments. The distance was either 20 or 15 cm, and the duration of treatment either 10, 20, 40 or 45 minutes.

As far as the general physiological effect of ultra violet treatment is concerned it was observed that the loss of fertility was proportionally greater than in the case of centrifugation. The lethal effect on the parent was also stronger. Generally speaking, treatment of

20 minutes or less at a distance of 20 cm or longer had no great lethal effect, but when treatment lasted for 40 minutes at a distance of 15 cm or less it was found that the duration of life of the parent treated was less than 12 days or little more than 6 days.

A summary of the results obtained for the repulsion experiment is given in the following table.

TABLE 20. Repulsion b.c. General results obtained for ultra violet treatment of $F_1$ females.

| culture | 1st bottles | | | | 2nd bottles | | | | 3rd bottles | | | |
|---|---|---|---|---|---|---|---|---|---|---|---|---|
| | bV | Bv | BV | bv | bV | Bv | BV | bv | bV | Bv | BV | bv |
| 112 | 49 | 51 | 12 | 12 | 62 | 54 | 14 | 12 | | | | |
| 122 | 88 | 89 | 26 | 27 | 40 | 46 | 16 | 21 | 42 | 45 | 11 | 17 |
| 123 | 45 | 36 | 12 | 14 | 66 | 75 | 24 | 27 | 53 | 39 | 21 | 18 |
| 124 | 68 | 45 | 14 | 8 | 40 | 41 | 11 | 11 | | | | |
| 126 | 31 | 42 | 9 | 20 | 50 | 32 | 16 | 14 | 3 | 2 | 0 | 2 |
| 127 | 38 | 42 | 10 | 10 | 81 | 74 | 11 | 12 | | | | |
| 128 | 57 | 34 | 14 | 9 | 110 | 122 | 24 | 24 | | | | |
| 129 | 94 | 103 | 15 | 19 | 34 | 32 | 3 | 8 | 51 | 58 | 13 | 13 |
| total | 471 | 442 | 112 | 119 | 483 | 476 | 119 | 129 | 149 | 114 | 45 | 50 |
| c.o.v. | 231/1144 = 20.19% | | | | 248/1207 = 20.54% | | | | 95/388 = 24.5 % | | | |
| M c.o.v. | 20.99, m ± 0.54 | | | | 21, m ± 2 | | | | 24.79, m ± 2.2 | | | |

TABLE 21. Repulsion b.c. Effect of duration of treatment of $F_1$ females at 20 cm distance.

| duration | culture | M c.o.v. 1st bottles | M c.o.v. 2nd bottles | M c.o.v. 3rd bottles |
|---|---|---|---|---|
| 10 min. | 112 | 19.35 | 10.01 | |
| | 124 | 16.3 | 21.36 | |
| | c.o.v. of totals | 17.72 | 20.88 | |
| 20 min. | 126 | 28.43 | 26.78 | |
| | 127 | 20.0 | 12.92 | |
| | c.o.v. of totals | 23.16 | 18.27 | |

TABLE 22. Repulsion b.c. Effect of duration of treatment of $F_1$ pupae at 20 cm distance.

| duration | culture | M c.o.v. 1st bottles | M c.o.v. 2nd bottles | M c.o.v. 3rd bottles |
|---|---|---|---|---|
| 10 min. | 122 | 23.04 | 30.1 | 24.35 |
| | 128 | 20.18 | 17.14 | |
| | 129 | 14.66 | 14.28 | 19.26 |
| | c.o.v. of totals | 19.1 | 20.0 | 21.6 |

TABLE 23. Repulsion b.c. Effect of duration of treatment of $F_1$ larva at 20 cm distance.

| duration | culture· | c.o.v. 1st bottles | c.o.v. 2nd bottles | c.o.v. 3rd bottles |
|---|---|---|---|---|
| 10 min. | 123 | 25.69 | 26.56 | 29.77 |

From these tables it appears that the flies treated for 20 minutes at a distance of 20 cm show a higher c.o.v. for the 1st bottles than the c.o.v. of the 1st bottles of the flies treated for 10 minutes. Pupae treated for 10 minutes at 20 cm showed an effect intermediate between that of flies treated for 10 minutes and flies treated for 20 minutes, while the larva treated for 10 minutes also at 20 cm, showed the strongest effect. As far as the 2nd and the 3rd bottles are concerned the grand totals resp. of flies, pupae and larva treated for 10 or 20 minutes show a higher c.o.v. than the controls for the repulsion experiments, and this phenomenon clearly displays itself in the respective grand totals of the 1st, 2nd and 3rd bottles of flies, pupae and larva, as follows:

TABLE 24.  Repulsion b.c. Comparison of c.o.v. of treated with
controls, 1. M c.o.v. % 2, c.o.v. % of absol. total.

| | M c.o.v. 1st bottles | M c.o.v. 2nd bottles | M c.o.v. 3rd bottles |
|---|---|---|---|
| 1.treated<br>contr. | 20.99, m 0.54<br>20.44, m 0.4 | 20.66, m 2.0<br>17.54, m 1.64 | 24.79, m $\pm$ 2.2<br>13.8 |
| diff. | 0.55, m diff. $\pm$0.67 | 3.12, m diff. $\pm$2.59 | 10.99 |
| | c.o.v. of absolute<br>total | c.o.v. of absolute<br>total | c.o.v. of absolute<br>total |
| 2. treated<br>contr. | 20.19 %<br>19.8  % | 20.54 %<br>17.84 % | 24.51 %<br>13.8  % |
| diff. | 0.39 % | 2.70 % | 10.71 % |

Also the grand totals for the repulsion experiments in connection
with ultra violet treatment may be compared with the grand totals
of the repulsion controls as follows.

TABLE 25.  Repulsion b.c. Grand totals of ultra violet light and
controls compared.

| | bV | Bv | BV | bv | c.o.v. % |
|---|---|---|---|---|---|
| treated | 1329 | 1314 | 347 | 392 | 21.85 |
| controls | 443 | 390 | 97 | 106 | 19.59 |

Neither for the 1st nor for the 2nd bottles is the difference between
the c.o.v.'s of the treated and the controls statistically significant,
the difference being always less than 3 $\times$ m diff. On the other hand
however, 14 out of 20 cultures for the 1st, 2nd and 3rd bottles show
a higher c.o.v. for the treated as compared with the controls, 2 cultures had more or less the same values, while 4 values were lower. We
are justified, I think, to conclude that as regards the repulsion experiments, ultra violet treatment raised the percentage of crossing-
over between black and vestigial.

TABLE 26. Coupling b.c. General results obtained for ultra violet treatment of $F_1$ females.

| culture | 1st bottles | | | | 2nd bottles | | | | 3rd bottles | | | |
|---|---|---|---|---|---|---|---|---|---|---|---|---|
| | bV | Bv | BV | bv | bV | Bv | BV | bv | bV | Bv | BV | bv |
| 145 | 6 | 4 | 31 | 31 | 7 | 5 | 37 | 36 | | | | |
| 146 | 9 | 6 | 67 | 70 | | | | | | | | |
| 147 | 7 | 6 | 35 | 37 | | | | | | | | |
| 148 | 4 | 2 | 16 | 26 | | | | | | | | |
| 152 | 1 | 2 | 32 | 18 | 5 | 7 | 77 | 62 | 6 | 10 | 62 | 52 |
| 153a | 5 | 6 | 26 | 34 | | | | | | | | |
| 153b | 6 | 8 | 31 | 41 | | | | | | | | |
| 154 | 2 | 2 | 21 | 18 | | | | | | | | |
| 155 | | | 6 | 4 | 5 | 8 | 11 | 23 | 2 | 1 | 15 | 18 . |
| 156 | 3 | 5 | 42 | 44 | 5 | 5 | 18 | 21 | 4 | 5 | 15 | 17 |
| totals | 43 | 41 | 301 | 319 | 22 | 25 | 143 | 142 | 12 | 16 | 92 | 87 |
| c.o.v. | 84/704 = 11.93 % | | | | 47/332 = 14.16 % | | | | 28/207 = 13.53 % | | | |
| M c.o.v. | 12.27, m ± 0.96 | | | | 15.62, m ± 1.34 | | | | 14.18, m ± 3.2 | | | |

As to the coupling experiments with ultra violet treatment the following results were obtained.

The crossover values for the controls (coupling) were as follows.

TABLE. 27. Coupling b.c. c.o.v.'s for controls.

| culture | 1st bottles | | | | 2nd bottles | | | | 3rd bottles | | | |
|---|---|---|---|---|---|---|---|---|---|---|---|---|
| | bV | Bv | BV | bv | bV | Bv | BV | bv | bV | Bv | BV | bv |
| 136 | 6 | 7 | 33 | 40 | 6 | 6 | 43 | 59 | 4 | 6 | 27 | 26 |
| 144 | 7 | 6 | 47 | 34 | 6 | 7 | 24 | 28 | | | | |
| total | 13 | 13 | 80 | 74 | 12 | 13 | 67 | 87 | 4 | 6 | 27 | 27 |
| c.o.v. | 26/180 = 14.4 % | | | | 25/179 = 13.97 % | | | | 10/63 = 15.9 % | | | |
| M c.o.v. | 14.47, m ± 0.43 | | | | 15.35, m ± 3.28 | | | | | | | |

The crossover value for the 1st bottles of the treated flies was lower than that of the controls by 2.2 units, m diff. being 1.05. For the 2nd bottles there was a very small statistically insignificant difference between the c.o.v. of the treated and of the controls, the difference however being only a fraction of a unit in favour of the c.o.v. of the treated. But there was a difference of more than 3 units between the 1st and the 2nd bottles of the treated flies, the difference which was

in favour of the 2nd bottles was more than twice m diff. This phenomenon was apparent from the grand total of the 1st and the 2nd
bottles of the treated flies, as is apparent from table 26. For the controls
the difference between the 1st and the 2nd bottles was very small
or even doubtful. See table 27. It is difficult to interpret these results
and it is probably better to leave this matter of the effect of ultra violet
treatment on the c.o.v. of the different age period of the flies for
further investigation, and now to study the question whether for the
coupling experiment ultra violet light had an influence on
crossing-over at all and if so, whether we can detect a probable correlation between intensity and duration of treatment on the one hand
and the c.o.v. on the other.

As far as the grand totals are concerned the following were obtained.

TABLE 28.  Coupling b.c. totals; c.o.v. of ultra violet treated flies
and of controls compared.

|              | bV | Bv | BV  | bv  | c.o.v. |
|--------------|----|----|-----|-----|--------|
| ultra viol.  | 62 | 58 | 433 | 438 | 12.1   |
| controls     | 32 | 38 | 216 | 226 | 13.67  |

From this table it is apparent that the sub-classes were very well
represented so that the reliability of the c.o.v.'s is certainly good.
There was an absolute difference of 1.57 units i.e. an increase of the
controls of more than 13 %. It may be concluded therefore that for
the coupling back-crosses ultra violet treatment caused a decrease in
the percentage of crossing-over between black and vestigial.

We have thus observed a radical difference of the same physical
agent as to its effect on a coupling and a repulsion experiment in so
far as we have observed a positive increase in the c.o.v. for the ultra
violet repulsion experiment, while for the coupling experiments a
slight decrease in the c.o.v. was observed.

As regards the possible effects of different doses of ultra violet
treatment the following results were obtained.

TABLE 29. Coupling b.c. Effect of different doses ultra violet light amperes $\times$ min./cms².

| culture | stage when treated | relative strength of dose | c.o.v. % 1st bottles | c.o.v. % 2nd bottles | diff. 1st-2nd bottl. |
|---------|-------------------|--------------------------|---------------------|---------------------|---------------------|
| 152 | pupa | 0.154 | 9.3 | 8.62 | +0.68 |
| 145, 146, 147, 148 | 1 day old flies | 0.175 | 12.32 | 14.12 | —1.8 |
| 156 | pupa; fly 1 day old | 0.39 | 8.5 | 24.2 | —15.7 |
| 153ᵃ, 153ᵇ, 154 | pupa; fly 1 day old | 0.63 | 13.7 | | |

From the above table the difference between the 1st and 2nd bottles appear to increase proportional to the strength of the dose. But the sub-classes of cult. 152 are ill-represented, so that the table may be somewhat misleading as it stands.

### § 3. *Coupling and repulsion crossing-over values compared*

The difference in effect of the same physical agent in the case of a coupling and a repulsion experiment has probably a deeper meaning. For, as will be pointed out, the figures obtained so far showed a decided higher c.o.v. in the case of repulsion experiments as compared with coupling experiments. In the first instance we may compare the grand totals of all the coupling experiments with those of all the repulsion experiments.

TABLE 30. C.o.v.'s of repulsion experiments compared with c.o.v.'s of coupling experiments.

| | bV | Bv | BV | bv | c.o.v. % | M. c.o.v. % |
|---|-----|-----|------|------|----------|-------------|
| coupling | 260 | 256 | 1593 | 1614 | 13.86 | 13.3, m ± 0.6 |
| repulsion | 2337 | 2275 | 596 | 663 | 21.44 | 21.08, m ± 0.35 |

The difference between the repulsion c.o.v. and the coupling c.o.v. was 7.78, m diff. is 0.69 so that the difference is more than 11 $\times$ m diff.

From the literature on *Drosophila* it is difficult to see what really is the standard c.o.v. between black and vestigial. Great variations display themselves in the summary given by BRIDGES and MORGAN in publ. 278, "CARNEGIE Inst. of Washington", 1919, p. 154. If we calculate the mean c.o.v. from this table a percentage of 17.8, m ± 1.0 is obtained for the totals of repulsion + coupling experiments. If we balance our totals for control repulsion and control coupling a value of 16.63 % is obtained which agrees fairly well with BRIDGES' B.C. December 1913, with a c.o.v. of 16.5 %. It will further be observed that MORGAN obtained 18.1 % for his repulsion b.c. involving only black and vestigial, while he obtained 17.1 % for his coupling experiments involving the same two mutant factors. The latter amount differs from the former by 1 unit of crossing-over; in my coupling experiments were involved 6 mutant factors in the same chromosome and I obtained 5.9 units of crossing-over less than in the case of my repulsion experiments where there was one mutant factor in each partner chromosome, from which it appears that the difference in crossing-over between a repulsion and coupling experiment is roughly equal to the difference between the number of mutant factors concerned in the homologous chromosomes. This proportional difference therefore suggests a cytological explanation instead of the physiological explanation e.g. of difference in class viability. That which was perhaps looked upon as a handicap in the construction of chromosome maps, may perhaps prove itself to be a very serviceable instrument.

In this dissertation the following cases have bearing on the same problem.

BRIDGES (17) found in connection with an age experiment on crossing-over in *Drosophila melanogaster* for the cross wild × purple vestigial (coupling) for the 1st bottles a c.o.v. of 10.7; for the 2nd bottles a c.o.v. of 7.8. For the repulsion back-crosses he obtained for the 1st bottles a c.o.v. of 13, for the 2nd bottles a c.o.v. of 8.1. This means that he obtained for his coupling back-crosses a mean c.o.v. of 9.2 and for his repulsion back-crosses a mean c.o.v. of 10.5. STURTEVANT (112) reported the case of the crossover modifier XIII located somewhere to the left of purple which reduces crossing-over between Star and purple in females heterozygous for it; CIIr located between purple and speck reduces crossing-over in that region when it is present in heterozygous condition. These may in fact be explained as inverted chromosome sections. PAYNE (87) observed that when his 9-bristle stock was crossed to a multiple recessive stock, crossing-over was greatly reduced in the $F_1$ females. PAYNE considered this to be due to a gene in the 3rd chromosome which was non-lethal in homozygous condition. BRIDGES called this modifier CIIIP. STURTE-VANT reported the crossover modifier CIII which when present in heterozygous condition greatly reduced crossing-over, but crossing-over remained normal when CIII was present in homozygous condition. SEREBROVSKY (100) obtained a reduction in crossing-over between black and cinnabar when pp was substituted for PP. Reduction was still more marked when Pp was substituted for PP. SEREBROVSKY, IVANOVA and FERRY (101) reported the case of the influence of y, li and Ni on crossing-over close to their loci in the sex-chromosome. Reduction in crossing-over was greatly reduced in the case of coupling back-crosses. COLLINS and KEMPTON (25) found in connection with linkage of C and Wx in maize that crossing-over was very much lower when R was present in heterozygous condition than when it was present in homozygous condition. Crossing-over between C and Wx was also reduced when for SuSu was substituted Susu. HUXLEY (60) found in connection with *Gammarus chevreuxy* much lower c.o.v.'s for the coupling back-cross BCbc × bbcc than for his repulsion back-cross BcbC × bbcc. DETLEFSEN (30) found for his studies on mice crossing-over between eye colour and body colour to be twice as frequent for repulsion back-crosses as for coupling back-crosses.

Summarising, centrifugation generally caused a rise in crossing-over for both repulsion and coupling experiments; ultra violet light caused a rise in crossing-over for repulsion but a decrease in crossing-over for coupling experiments. Crossover value obtained from a repulsion experiment was found higher than that of the coupling, roughly proportional to the difference in number of mutant factors in the homologous chromosomes, a difference of n mutant factors between the homologous chromosomes causing a difference of n units of crossing-over. This probably is in keeping with the greater or smaller symmetry between the homologous chromosomes. This fact therefore is strongly supported by SEREBROVSKY's findings discussed' on p. 24 of this dissertation, but I intend systematically to test out the suggestion which I made here and elsewhere in this dissertation viz. p. 25. The c.o.v.'s of repulsion and coupling experiments will be compared for as many cases as possible of asymetry between the homologous chromosomes.

Summary

Part I

A study of the literature was made as to the changes in the crossing-over value and it was found that these may be brought about by physical agencies e.g. temperature X-rays and radium; by physiological agencies e.g. age and sex; by genetical causes e.g. crossover modifiers. Crossing-over was studied in cases of non-disjunction and polyploidy, of translocation, deletion and deficiency. In certain cases in fish the frequency of crossing-over from X to Y was compared with that from Y to X. The outcome of these studies may be briefly summarised.

1. Inter- and intra-chromosomal differences in reaction to the same physical agent was found. The central regions of the 2nd and the 3rd chromosome and the right end portion of the 1st chromosome of *Drosophila melanogaster* were found to be more liable to changes in the c.o.v. than other regions. These sensitive regions correspond with the place of attachment of the spindle-fibre.

2. Studies in cases of non-disjunction and polyploidy showed that the Y-chromosome had no effect on crossing-over in XXY females; crossing-over may take place between three homologues of a triploid female.

3. Very important were the results obtained in connection with translocation, deletion, inversion and deficiency. These studies suggest a new line of action, which, with due attention to the foresaid factors, makes it possible to construct chromosome maps which are more in agreement with the actual chromosome as has been hitherto the case. Translocations have the advantage that they may be evoked experimentally by X-rays.


Part II

Experiments were done with *Lebistes reticulatus* in order to obtain data which will make it possible to compare the crossing-over value when crossing-over takes place from X to Y with that obtained when crossing-over takes place from Y to X. The problem at issue is that the difference in frequency between the two directions of crossing-over suggest an exchange adaptation in so far as the recipient chromosome has to adapt itself to the donor chromosome. The results were briefly as follows.

1. Crossing-over of the elongatus (El) factor from Y to X takes place in 1.8 percent of the cases.

2. Results obtained from the cross $X_oX_o \times X_{El}Y_{Ma}$ in order to determine the frequency of crossing-over of El from X to Y were quite contrary to expectation, for, with the exception of 3 males with a round caudal fin, all others had an elongated caudal fin. This suggests that the parent male was not of the constitution $X_{El}Y_{Ma}$, neither could it have been of the constitution $X_{El}Y_{Ma,El}$, so that this remains to be made out by suitable experiments. So far crossing-over from X to Y takes place in $\pm$ 1.4 percent of the cases according to WINGE.

3. The factor for maculatus which has hitherto been looked upon as sex-limited, has been found to cross over from Y to X.

4. The expression of the El factor was studied in connection with the development of the colour pattern of the male *Lebistes*. It was found to express itself relatively later than most of the main colour patterns.

5. Technical suggestions were made.

Experiments were also done with *Drosophila melanogaster* and the effects of centrifugation and of treatment with ultra violet light were studied on crossing-over between black and vestigial in the second chromosome. Special study was made of the difference in crossing-over between repulsion and coupling back-crosses. The results were as follows.

1. Centrifugation caused an increase in the crossing-over value for the repulsion as well as for the coupling back-crosses.

2. Treatment with ultra violet light caused an increase in the crossing-over value for repulsion but a slight decrease for the coupling back-crosses.

3. The crossing-over value for repulsion back-crosses for all experiments was much higher than that for coupling back-crosses, the totals were very reliable and the difference between the crossing-over value for repulsion and the crossing-over value for coupling was found

to be 11 × m diff. A brief survey of cases discussed in part 1 together with own data show that the difference was always in favour of the repulsion back-crosses. It is suggested that the difference is roughly proportional to the difference between the number of mutant factors concerned in the homologues. It is further suggested that the difference is not due to the differences in the viability of the classes, but to the relatively greater asymetry between the homologues in the case of a coupling back-cross.

# LITERATURE

1. AIDA, T., On the inheritance of colour in a fresh water fish *Aplocheilus latipes* TEMMINCK and SCHLEGEL, with special reference to sex-linked inheritance. Genetics 6, 1921, p. 554.
2. AIDA, T., Further genetical studies on *Aplocheilus latipes*. Genetics 16, 1930, p. 1.
3. ALBERTS, H. W., A method for calculating linkage values. Genetics 11, 1926, p. 235.
4. ANDERSON, E. G., Crossing-over in a case of attached X-chromosomes in *Drosophila melanogaster*. Genetics 10, 1925, p. 403.
5. ANDERSON, E. G., Studies on a case of high non-disjunction in *Drosophila melanogaster*. Zeitschr. f. ind. Abst. und Vererb. 51, 1929, p. 397.
6. ALTENBURG, E., Linkage in *Primula sinensis*. Genetics 1, 1916, p. 354.
7. BATESON, W., MENDEL's principles of heredity. 1913.
8. BAUR, E., Einführung in die Vererbungslehre, Berlin, 1930.
9. BÊLÁR, K., Neuere Untersuchungen über Geschlechtschromosomen bei Pflanzen. Sammelreferat. Zeitschr. f. ind. Abst. u. Vererb. 35, 1924, p. 172.
10. BERGNER, A. D., The effect of prolongation of the lifecycle on crossing-over in the second and the third chromosomes of *Drosophila melanogaster*. Journ. of Exp. Zool. 50, 1928, p. 107.
11. BLACHER, L. J., The dependance of secondary sex characters upon testicular hormones in *Lebistes reticulatus*, Biol. Bull. 50, 1926, p. 374.
12. BLACHER, L. J., Materials for the genetics of *Lebistes reticulatus*. Transactions of the Lab. of Exp. Biol. of the Zoopark of Moscow 4, 1928, p. 245.
13. BOLEN, H. R., A mutual translocation involving the fourth and the X-chromosomes of *Drosophila*. Am. Nat. 65, 1931, p. 417.
14. BREGGER, T. J., Linkage in maize; the C aleurone factor and waxy endosperm. Am. Nat. 52, 1918.
15. BRIDGES, C. B., A linkage variation in *Drosophila*. Journ. of Exp. Zool. 19, 1915, p. 1.
16. BRIDGES, C. B., Deficiency. Genetics 2, 1917, p. 445.
17. BRIDGES, C. B., Variations in crossing-over in relation to age of female in *Drosophila melanogaster*. Carnegie Inst. of Washington, Publ. 399, 1929, p. 63.
18. BRIDGES, C. B. and ANDERSON, E. G., Crossing-over in the X-chromosomes of triploid females of *Drosophila melanogaster*. Genetics 10, 1925, p. 418.
19. BRIDGES, C. B. and MORGAN, T. H., The second chromosome group of mutant characters. Carnegie Inst. of Washington, Publ. 278, 1919, p. 125.

20. CASTLE, W. E., A sex difference in linkage in rats and mice. Genetics 10, 1925, p. 580.

21. CASTLE, W. E., Contributions to a knowledge of inheritance in mammals. 1926, Part 1, p. 16.

22. CASTLE, W. E., and WACHTER, W. L., Variations of linkage in rats and mice. Genetics 9, 1924, p. 1.

23. COLE, L. J., A case of sex-linked inheritance in the domestic pigeon. Science 36, 1912, p. 190.

24. COLE, L. J. and KELLY, F. J., Studies on inheritance in pigeons 3. Description and linkage relations of two sex-linked characters. Genetics 4, 1919, p. 183.

25. COLLINS, G. N. and KEMPTON, J. H., Variability in the linkage of two seed characters of maize. Bull. U.S. Dept. Agric. 1468, 1926.

26. CHRISTIE, W. and WRIEDT, C., Die Vererbung von Zeichnungen, Farben und anderen Characteren bei Tauben. Zeitschr. f. ind. Abst. u. Vererb. 32, 1924, p. 233.

27. DAVENPORT, C. B., Sex-limited inheritance in poultry. Journ. of Exp. Zool. 13, 1912, p. 1.

28. DEMEREC, M., A possible explanation for WINGE's findings in *Lebistes reticulatus*. Am. Nat. 62, 1928, p. 90.

29. DETLEFSEN, J. A., Is crossing-over a function of distance? Proc. of Nat. Acad. of Sciences (U.S.A.) 6, 1920, p. 663.

30. DETLEFSEN, J. A., The linkage of dark eye and color in mice. Genetics 10, 1925, p. 17.

31. DETLEFSEN, J. A. and CLEMENTE, L. S., Linkage of a dilute color factor and dark eye in mice. Genetics 9, 1924, p. 247.

32. DETLEFSEN, J. A. and ROBERTS, E., Studies on crossing-over. 1. The effect of selection on crossing-over values. Journ. of Exp. Zool. 32, 1921, p. 333.

33. DOBZHANSKY, TH., Translocations involving the third and the fourth chromosomes of *Drosophila melanogaster*. Genetics 15, 1930, p. 347.

34. DOBZHANSKY, TH., The decrease of crossing-over observed in translocations and its probable explanation. Am. Nat. 65, 1931, p. 214.

35. DOBZHANSKY, TH., Studies on chromosome conjugation. 1. Translocation involving the second and Y-chromosome of *Drosophila melanogaster*. Zeitschr. f. ind. Abst. u. Vererb. 60, 1932, p. 235.

36. DUNN, L. C., Linkage in mice and rats. Genetics 5, 1920, p. 325.

37. EMERSON, R. A., The calculation of linkage intensities. Am. Nat. 50, 1916, p. 411.

38. EMERSON, R. A. and HUTCHINGON, C. B., The relative frequency of crossing-over in micro- and megaspore development in maize. Genetics 6, 1921, p. 417.

39. EYSTER, W. H., The intensity of linkage between the factors for sugary endosperm and for truncate ears, and the relative frequency of their crossing-over in microspore and megaspore development (in maize). Genetics 7, 1922, p. 597.

40. EYSTER, W. H., The linkage relations between the factors tunicate ear starchy sugary endosperm in *maize*. Genetics 6, 1921, p. 209.
41. EYSTER, W. H., Five new genes in chromosome 1 in maize. Zeitschr. f. ind. Abst. u. Vererb. 49, 1929, p. 105.
42. FISHER, R. A., The systematic location of genes by means of crossing-over observations. Am. Nat. 56, 1922, p. 406.
43. FRASER, A.C. and GORDON, M., Crossing-over between the W- and Z-chromosomes of the killifish *Platypoecilus*. Science N.S. 67, N° 1740, 1928, p. 470.
44. FRASER, A. C. and GORDON, M., The genetics of *Platypoecilus*. II. The linkage of two sex-linked characters. Genetics 14, 1929, p. 160.
45. GATES, W. H., Linkage of the factors for short ear and density in the house mouse (*Mus musculus*). Genetics 13, 1928, p. 170.
46. GOLDSCHMIDT, R., Einführung in die Vererbungswissenschaft. 1928.
47. GOODALE, H. D., Crossing-over in the sex-chromosomes of the male fowl. Science. N.S. 46, 1909, p. 232.
48. GOWEN, J. W., A biometrical study of crossing-over. Genetics 4, 1919, p. 205.
49. GOWEN, J. W., The cell division at which crossing-over takes place. Proc. Nat. Acad. of Science 15, 1929, p. 266.
50. GOWEN, J. W. and W. S., Complete linkage in *Drosophila melanogaster*. Am. Nat. 56, 1922, p. 286.
51. GREGORY, R. P., On gametic coupling and repulsion in *Primula sinensis*. Proc. of the Roy. Soc. of London 84, N° 568, 1911.
52. GREGORY, R. P., DE WINTON, Miss D. and BATESON, W., Genetics of *Primula sinensis*. Journ. of Genetics 13, 1923, p. 219.
53. HALDANE, J. B. S., The combination of linkage values and the calculation of distance between the loci of linked factors. J. of Gen. 8, 1919, p. 299.
54. HALDANE, J. B. S., The probable errors of calculated linkage values and most accurate method of determining gametic from certain zygotic series. Journ. of Genetics 8, 1919, p. 291.
55. HALDANE, J. B. S., Note on a case of linkage in *Paratettix*. Journ. of Genetics 10, 1920, p. 47.
56. HALDANE, J. B. S. and CREW, F. A. E., Change of linkage in poultry with age. Nature 115, 1925, p. 651.
57. HAMLETT, G. W. D., The linkage disturbance involved in the chromosome translocation I of *Drosophila*, and its probable significance. Biol. Bull. 51. 1926, p. 435.
58. HAMMARLUND, C. and HÅKANSSON, A., Parallelism of chromosome ring formation, sterility and linkage in *Pisum*. Hereditas 14, 1930, p. 97.
59. HERTWIG, PAULA., Tabellen der Vererbungslehre. Tabulae Biologicae 4, 1917, p. 114.
60. HUXLEY, J. S., Sexual difference of linkage in *Gammarus chevreuxi*. Journ. of Genetics 20, 1928, p. 145.

61. IMAI, Y., Linkage groups of the Japanese Morning Glory. Genetics 14, 1929, p. 223.
62. IMMER, F. R., Formulae and tables for calculating linkage intensities. Genetics 15, 1930, p. 81.
63. JUST, G., Die Konstanz der Faktorenaustauch-Werte. Biol. Zentralbl. 44, 1924, p. 258.
64. JUST, G., Untersuchungen über Faktorenaustausch. I. Untersuchungen zur Frage der Konstanz der crossing-over Werte. Zeitschr. f. ind. Abst. u. Vererb. 46, 1924, p. 95.
65. JOHANSSEN, W., Elemente der exakten Erblichkeitslehre. 1926.
66. KAPPERT, H., Ueber die Zahl der unabhängigen Merkmalsgruppen bei der Erbse. Zeitschr. f. ind. Abst. u. Vererb. 36, 1924, p. 1.
67. KÖRÖSY, K., Versuch einer Theory der Genkoppelung. Biblth. Genetica 15, 1929, p. 1.
68. KOSSWIG, C., Die Bedeutung des Y-Chromosoms im Tierreich. Der Züchter 2, 1930, p. 263.
69. MAKITA, K., Modifying factors influencing the striped-yellow crossing-over value in the Silkworm. (Japanese with summary in English). Bulteno Scienca de la Fakultato Terkultura, Kyushu Imp. Univ. 2, 1926, p. 33.
70. MAVOR, J. W., An effect of X-rays on the linkage of Mendelian characters in the first chromosome of *Drosophila*. Genetics 8, 1923, p. 355.
71. MAVOR, J. W., The effect on crossing-over and non-disjunction of X-raying the anterior and posterior halves of *Drosophila* pupae. Genetics 14, 1929, p. 129.
72. MAVOR, J. W. and SVENSON, H. K., Crossing-over in the second chromosome of *Drosophila melanogaster* in the $F_1$ generation of X-rayed females. Am. Nat. 58, 1924, p. 311.
73. MAVOR, J. W. and SVENSON, H. K., A comparison of the effects of X-rays and temperature on linkage and fertility in *Drosophila*. Genetics 9, 1924, p. 588.
74. MAVOR, J. W. and SVENSON, H. K., An effect of X-rays on the linkage of Mendelian characters in the second chromosomes of *Drosophila melanogaster*. Genetics 9, 1924, p. 70.
75. METZ, C. W. and MOSES, M. S., A comparison of the chromosomes of different species of *Drosophila*. Journ. of Heredity 14, 1923, p. 195.
76. MOHR, O. L., A genetical and cytological analysis of a section deficiency involving four units of the X-chromosome in *Drosophila* melanogaster. Zeitschr. f. ind. Abst. u. Vererb. 32, 1923, p. 108.
77. MORGAN, L. V., Polyploidy in *Drosophila melanogaster* with two attached chromosomes. Genetics 10, 1925, p. 148.
78. MORGAN, T. H., The physical basis of heredity. 1919.
79. MULLER, H. J., The regionally differential effect of X-rays on crossing-over in autosomes of *Drosophila*. Genetics 10, 1925, p. 470.

80. MULLER, H. J., Induced crossing-over variation in the X-chromosome of *Drosophila*. Am. Nat. 60, 1926, p. 192.
81. MULLER, H. J. and ALTENBURG, E., Chromosome translocations produced by X-rays in *Drosophila*. (Abstract) Anat. Rec. 41, 1928, p.100.
82. MULLER, H. J. and ALTENBURG, E., The frequency of translocations produced by X-rays in *Drosophila*. Genetics 15,1930, p. 283.
83. MULLER, H. J. and PAINTER, T. S., The cytological expression of changes in gene alignment produced by X-rays in *Drosophila*. Am. Nat. 63, 1928, p. 193.
84. NABOURS, R. K., Parthenogenesis and crossing-over in the grouse locust *Apotettix*. Am. Nat. 53, 1919, p. 131.
85. OWEN, F. V., Calculating linkage intensity by product moment correlation. Genetics 13, 1928, p. 80.
86. PAINTER, T. S. and MULLER, H. J., Parallel cytology and genetics of induced translocations and deletions in *Drosophila*. Journ. of Heredity 20, 1929, p. 287.
87. PAYNE, F., Crossover modifiers in the third chromosome of *Drosophila melanogaster*. Genetics 9, 1924, p. 327.
88. PLOUGH, H. H., The effect of temperature on crossing-over. Journ. of Exp. Zool. 24, 1917, p. 147.
89. PLOUGH, H. H., The effect of temperature on linkage in the second chromosome of *Drosophila*. Proc. Nat. Acad. Sc. 3, 1917.
90. PLOUGH, H. H., Further studies on the effect of temperature on crossing-over. Journ. of Zool. 32, 1921, p. 187.
91. PLOUGH, H. H., Radium radiations and crossing-over. Am. Nat. 58, 1924, p. 85.
92. PUNNETT, R. C., Reduplication series in sweetpeas. Journ. of Genet. 3, 1913, p. 77.
93. RASMUSSON, J., Genetically changed linkage values in *Pisum*. Heriditas 10, 1927—28, p. 1.
94. REDFIELD, H., Crossing-over in the third chromosomes of triploids of *Drosophila melanogaster*. Genetics 15, 1930, p. 205.
95. RHOADES, M. M., A new type of translocation in *Drosophila melanogaster*. Genetics 16, 1931, p. 490.
96. ROBERTSON, W. R. B., Chromosome studies III. Inequalities and deficiencies in homologous chromosomes: their bearing upon synapsis and the loss of unit characters. Journ. of Morph. 26, 1915.
97. SCHMIDT, H., Geschlechtsumwandlungen bei tropischen Zierfischen. Der Züchter 2, 1930, p. 297.
98. SEILER, J., Die crossing-over Studien der Schule MORGAN. Naturw. 12, 1924, p. 677.
99. SEILER, J., Die Chiasmatypie als Ursache des Faktorenaustausches. Sammelreferat. Zeitschr. f. ind. Abst. und Vererb. 41, 1926, p. 259.
100. SEREBROVSKY, A. S., The influence of the purple gene on the crossing-over between black and cinnabar in *Drosophila melanogaster*. Journ. of Gen. 18, 1927, p. 137.

101. SEREBROVSKY, A. S., IVANOVA, O. A. and FERRY, L., On the influence of the genes y, li, and Ni on crossing-over close to their loci in the sex-chromosomes of *Drosophila melanogaster*. Journ. of Gen. 21, 1929, p. 287.
102. STADLER, L. J., The variability of crossing-over in maize. Genetics 11, 1926, p. 1.
103. SNELL, G., Inheritance in the house mouse, the linkage relations of short ear, hairless and naked. Genetics 16, 1931, p. 42.
104. STERN, C., Ein neue Chromosom Aberration von *Drosophila melanogaster* und ihre bedeutung für die Theorie der linearen Anordnung der Genen. Biol. Zentrbl. 46, 1926, p. 505.
105. STERN, C., An effect of temperature and age on crossing-over in the first chromosome of *Drosophila melanogaster*. Proc. Nat. Acad. of Sc. U.S.A. 12, 1926, p. 530.
106. STERN, C., Ein genetischer und zytologischer Beweis für Vererbung in Y-Chromosom von *Drosophila melanogaster*. Zeitschr. f. ind. Abst. u. Verb. 44, 1927, p. 188.
107. STERN, C., Untersuchungen über Aberrationen des Y-chromosoms von *Drosophila melanogaster*. Zeitschr. f. ind. Abst. u. Vererb., 51, 1929, p. 253.
108. STERN, C., Konversionstheorie und Austauschtheorie. Biol. Zentrbl. 50, 1930, p. 608.
109. STURTEVANT, A. H., An experiment dealing with sex-linkage in fowls. Journ. Exp. Zool. 12, 1912, p. 449.
110. STURTEVANT, A. H., The Linear arrangement of sex-linked factors in Drosophila. Journ. Exp. Zool. 14, 1913, p. 43.
111. STURTEVANT, A. H., Genetic factors affecting the strength of linkage in *Drosophila*. Proc. Nat. Acad. of Sc. 3, 1917, p. 555.
112. STURTEVANT, A. H., Inherited linkage variations in the second chromosome of *Drosophila*. Carn. Inst. Washington, Publ. 278, 1919, p. 305.
113. STURTEVANT, A. H., Genetic studies on *Drosophila simulans*. Genetics 6, 1921, p. 43.
114. STURTEVANT, A. H., A case of rearrangement of genes in *Drosophila*. Proc. Nat. Acad. of Sc. 7, 1921, p. 235.
115. STURTEVANT, A. H., The effects of unequal crossing-over at the bar locus in *Drosophila*. Genetics 10, 1925, p. 117.
116. STURTEVANT, A. H., A crossover reducer in *Drosophila melanogaster* due to inversion of a section of the third chromosome. Biol. Zentrbl. 46, 1927, p. 697.
117. STURTEVANT, A. H., The genetics of *Drosophila simulans*. Carn. Inst. of Washington, Publ. 399, 1929, p. 1.
118. STURTEVANT, A. H., BRIDGES, C. B. and MORGAN, T. H., The spatial relation of the genes. Proc. Nat. Acad. Sc., 5, 1919, p. 168.
119. STURTEVANT, A. H., and DOBZHANSKY, TH., Reciprocal transloca-

tions in *Drosophila* and their bearing on *Oenothera* cytology and genetics. Proc. Nat. Acad. Sci., 16, 1930, p. 533.

120. SVERDRUP, A., Linkage and independant inheritance in *Pisum sativum*. Journ. of Gen. 17, 1927, p. 221.

121. TANĀKA, Y., Occurrence of different systems of gametic reduplica-tion in male and female hybrids. Zeitschr. f. ind. Abst. u. Vererb. 14, 1915, p. 12.

122. TROW, A. H., A criticism of the hypothesis of crossing-over. Journ. of Gen. 5, 1916, p. 281.

123. VILMORIN DE, P. and BATESON, W., A case of genetic coupling in *Pisum*. Proc. Royal Soc., 84, 1911.

124. WACHTER, L., Data concerning linkage in mice. Am. Nat. 55, 1921, p. 412.

125. WARD, L., The genetics of curly wing in *Drosophila*. Genetics 8, 1923, p. 289.

126. WEINSTEIN, A., Coincidence of crossing-over in *Drosophila melano-gaster (ampelophila)*. Genetics 3, 1918. p. 135.

127. WEINSTEIN, A., The production of mutations and re-arrangements of genes by X-rays. Science 67, 1756, 1928.

128. WELLENSIEK, S. J., Genetic monograph on *Pisum*. Bibliogr. Ge-netica 2, 1925, p. 343.

129. WELLENSIEK, S. J., The occurrence of more than fifty percent crossing-over in *Pisum*. Genetica 11, 1929, p. 509.

130. WELLENSIEK, S. J., Linkage studies in *Pisum* III. Genetica 12, 1930, p. 1.

131. WINGE, Ö., A peculiar mode of inheritance and its cytological explanation. Journ. of Gen. 12, 1922, p. 137.

132. WINGE, Ö., Onesided masculine and sex-linked inheritance in *Lebistes reticulatus*. Journ. of Gen. 12, 1922, p. 145.

133. WINGE, Ö., On a Y-linked gene in *Melandrium*. Hereditas 9, 1922, p. 274.

134. WINGE, Ö., Crossing-over between the X- and Y-chromosome in *Lebistes*. Journ. of Gen. 13, 1923, p. 201.

135. WINGE, Ö., The location of eighteen genes in *Lebistes reticulatus*. Journ. of Gen. 18, 1927, p. 1.

136. WINGE, Ö., On the occurrence of XX males in *Lebistes* with some remarks on AIDA's socalled non-disjunctional males in *Aplocheilus*. Journ. of Gen. 23, 1930, p. 69.

137. WINKLER, H., Die Konversion der Gene. 1930.

138. WINKLER, H., Konversion-theorie und Austausch-theorie. Biol. Zentr.bl. 52, 1923, p. 164.

139. WINTON, DOROTHEA DE, Further linkage work on *Pisum sativum* and *Primula sinensis*. Zeitschr. f. ind. Abst. u. Vererb. Suppl. 2, 1928, p. 1594.